Supercomputing and the Transformation of Science

This picture, adapted from Katsushika Hokusai's masterpiece "The Great Wave off Kanagawa," artistically displays the spirit of supercomputing. Complex phenomena, such as waves on the surface of a fluid, are modeled by covering space with a grid and then solving the laws of physics at discrete points on that grid. The finer the grid, the closer the numerical simulation is to the actual solutions of the mathematical laws of nature that govern the physical world.

SUPERCOMPUTING AND THE TRANSFORMATION OF SCIENCE

William J. Kaufmann III
Larry L. Smarr

**SCIENTIFIC
AMERICAN
LIBRARY**

A division of HPHLP
New York

Library of Congress Cataloging-in-Publication Data

Kaufmann, William J.
    Supercomputing and the transformation of science / William J.
  Kaufmann III, Larry L. Smarr.
      p. cm.
    Includes bibliographical references (p.) and index.
    ISBN 0-7167-5038-4
    1. Supercomputers. 2. Science—Data processing. I. Smarr,
  Larry
L. II. Title.
QA76.88.K38 1993   92-32418
502′.85′411—dc20   CIP

ISSN 1040-3213

Printed in the United States of America

Scientific American Library
A division of HPHLP
New York

Distributed by W. H. Freeman and Company
41 Madison Avenue, New York, NY 10010
20 Beaumont Street, Oxford OX1 2NQ, England

2 3 4 5 6 7 8 9 0 KP 9 9 8 7 6 5 4 3

This book is number 43 in a series.

*To the intellectual giants who made
the digital world possible:
Charles Babbage, who created the computer;
John von Neumann, who taught us how to use it;
and Vannevar Bush, who foresaw how computers and
communications would transform science and society.*

Contents

Preface

A new information reality, parallel to but distinct from our well-known physical reality, is emerging. In this digital reality, bits take the place of the fundamental atoms of the physical world. Scientific instruments, from telescopes to microscopes to space probes, are extending our senses to enormous distances and to tiny scales, even to other worlds. Yet, instead of producing analog photographic plates that end up being stored in a vault, modern sensors record their discoveries in digital data banks accessible to researchers anywhere over computer networks. The mathematical laws of gravity, gas dynamics, and quantum mechanics—the products of generations of scientific thought—can now be solved by digital computers to create numerical representations of the physical world of orbiting planets, thunderstorms, and new drugs. These simulations offer scientists a new means of exploring nature. Finally, our analog modes of communication by voice, print, and video are gradually being replaced by digital modes. Ultimately most of human knowledge will be stored in a common digital library.

The worldwide acceptance of the personal computer has given every desktop a window into the world of digital knowledge. Fiber optic networks are creating intricate connections between millions of desktop computers and the relatively few supercomputers, the fastest computers that exist. These same networks tie this computational infrastructure into the vast archives of data maintained by industry, government, and scientific laboratories. This shadow universe of information, the "cyberspace" of science fiction novelist William Gibson, is rapidly increasing its reach. In *Supercomputing and the Transformation of Science,* we explore the supercomputers that are the central powerhouses of this information space. Our book focuses on three themes: the evolution of supercomputers, the methodologies for using them to simulate nature, and their transformation of virtually every field of science and engineering.

The supercomputers of today run almost one trillion times faster than the fastest computer of fifty years ago. For comparison consider another transformational technology—transportation. Fifty years ago the fastest mode of transportation had a speed limit of a few hundred miles per hour, whereas today's interplanetary spacecraft attain astonishing speeds of several hundred thousand miles per hour. We can now travel a thousand times faster than we could fifty years ago, yet this speedup is only one-billionth of the increase in speed achieved by supercomputers in the same time!

Because of their speed, supercomputers can perform huge numbers of computations in a brief span of time. It is that ability that enables these machines to create simulations of the natural world. Mathematics is capable of capturing the rich phenomenology of nature, and supercomputers are capable of employing this mathematics to generate the billions of numbers necessary to simulate the behavior of natural phenomena. The rise of computer graphics has allowed these vast mountains of numbers to be translated into visual imagery that is more intuitively accessible to human beings. Throughout the book, we have used visual images instead of traditional equations to capture the essence of a technical subject.

One of our most challenging tasks as authors was to convey the fundamental idea behind most of the techniques used to adapt the mathematical equations of theoretical science for use on supercomputers: that idea is to replace the continuous world of nature with a model of that world formed of discrete units. This can be accomplished by a variety of methods: we can approximate a fluid by dividing the space it flows through into a large number of small boxes, we can represent an engineering device by a finite set of subelements, or we can simulate a galaxy as a large number of gravitating particles. As we move from the more familiar world of classical physics into the strange world of quantum physics, we show how less intuitive approaches are used to make the complexity of quantum mechanical systems computationally tractable.

After clarifying the methodology of supercomputing, we take the reader on a comprehensive tour of its frontier applications. We start our voyage of discovery with the world of the quantum and end with the cosmos, following a course that parallels the organization of physical reality into a hierarchy of levels of increasing physical scale. In this hierarchy, atoms build molecules, which in turn build cells or bulk materials, which in turn build bodily organs or manufactured objects or geological structures. As our book progresses, we move up the hierarchy: thus, in between the extremes of the atom

and the whole universe, we explore the worlds of biology, engineering, and the environment.

The virtual worlds simulated by supercomputers can be probed and measured much more thoroughly than the physical world. Our exploration of these worlds is already leading to deeper understandings, and it is also turning the supercomputer into an instrument of engineering design. The result is the creation of safer, cheaper, and more reliable products in less time than would otherwise be possible. Following its adoption a decade ago in the petroleum, automobile, and aerospace industries, supercomputing is now radically altering industries producing chemicals, pharmaceuticals, consumer products, and financial services. Furthermore, supercomputers are beginning to offer one of the few techniques for rationally analyzing the problems confronting our environment. Pollution, exhaustion of natural resources, ozone depletion, and global warming are but a few of the research frontiers explored by supercomputing.

As we approach the next century, we cannot help but wonder how supercomputing will transform our ability to manipulate the physical world. Simulations are giving us detailed knowledge of materials and of biological molecules. As we acquire this knowledge, we will undoubtedly be able to restructure the traditional substances and organisms found in nature today. As we learn how to digitally control the engineering process from design through manufacturing, we will radically enhance our abilities to convert basic materials into finished products. As we gain detailed and verified models of the ecological and environmental systems of the Earth, we will have vastly greater powers for altering our world. Our hope is that the ability of computers to assist human beings in understanding the complexities of our world will lead to a growth of wisdom adequate to the challenges created by these new technologies.

We would like to thank the thousands of researchers whose pioneering use of supercomputers inspired us to write this book. We have been able to include the work of only a few representatives from each discipline, and we apologize to all those whose research we were unable to cover. The researchers whose work we do describe often sent us materials or read drafts for accuracy on very short notice. We would like to single out those who took on the extra effort of reviewing entire sections in order to minimize the misrepresentations that are unavoidable in a broad survey such as this one. These people include Fouad Ahmad, David Ceperley, Robert Chervin, Art Freeman, Bruce Hannon, Michael Heath, Michael Fainan, Karl Hess, Eric Jakobsson, Radha Nandkumar, Michael Norman, David Pines, Michael Schlesinger, Harrell Sellers, Shankar Subramaniam, Robert Sugar, Warren Washington, Robert Wilhelmson, and Carl Woese. Finally, Michael Norman and Robert Wilhelmson put in extra effort to create original images and illustrations for our book.

Larry Smarr also benefitted from the generous help of a number of his fellow directors of supercomputing facilities, including Bill Buzzbee, Sid Karin, Mal Kalos, Michael Levine, Ralph Roskies, and Vic Peterson. The documentation staffs of these centers and the science writers for *Cray Channels* and *Supercomputing Review* made our job of identifying and developing stories about individual research efforts much easier than it would have been otherwise.

Bill Kaufmann, who traveled to numerous conferences, symposia, and supercomputing facilities throughout the world to gather material for this book, would like to thank his many friends and colleagues for their help and hospitality. Bill is especially grateful to the staffs at Los Alamos National Laboratory, Lawrence Livermore National Laboratory, NASA Ames Research Center, the Minnesota Supercomputing Institute, Sandia National Laboratory, the National Center for Supercomputing Applications, the National Center for Atmospheric Research, the European Centre for Medium-Range Weather Forecasts, the San Diego Supercomputing Center, and Cray Research, Inc.

The manuscript for this book demanded elaborate preparation, made necessary by the enormous scope of the subject matter. The support of the staff of the National Center for Supercomputing Applications, particularly in Applications, Documentation, and Media Services, was critical in many ways, as was NCSA's support structure provided by the National Science Foundation, the State of Illinois, the University of Illinois at Urbana-Champaign, and NCSA's corporate sponsors. Jim Bottum, Deputy Director of NCSA, kept the center running while the Director was busy writing. The Director's assistant, Janus Wehmer, and her secretary, Linda Griffet, provided support above and beyond the call of duty; without their dedicated efforts, the publication schedule for this book could never have been met.

It has taken five years to bring this book to fruition. During that time, the staff of the Scientific American Library has been very patient and enormously competent. From the beginning, publisher Jerry Lyons provided continual encouragement. Once we had completed our manuscript, Susan Moran provided meticulous and invaluable editing. Indeed, she deserves to be a third author of this book!

The Scientific American Library production staff did a wonderful job, considering the hundreds of color images that had to be collected and processed. We would especially like to acknowledge Larry Marcus for photo research, John Hatzakis for page layout, Alice Fernandes-Brown for our book's design, Tina Hastings for her work as project editor, Christine McAuliffe for her oversight of the line illustrations, and Sheila Anderson for her coordination of typesetting and printing.

Finally, Larry Smarr would like to express his deep appreciation for the unending patience of his wife, Janet, and his boys, Joseph and Benjamin, during the final period of preparation of the manuscript.

William J. Kaufmann III
Larry L. Smarr
September 1992

Supercomputing and the Transformation of Science

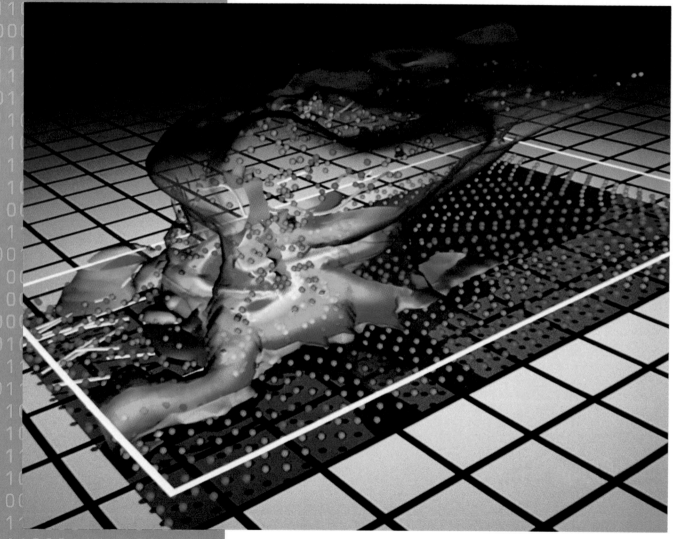

The Emergence of a Digital Science

Computers have a significant impact on daily life. They keep track of our telephone calls, compute our income taxes, and tally our charge cards. But one kind of computer, the supercomputer, is actually changing the way in which we conduct scientific research and engineering design. These computers, the most powerful that can be built, allow us to replace the physical world with a digital reality in the form of an array of numbers. When transformed into images on a computer screen, these numbers are easily seen to replicate essential features of phenomena in the natural world.

In this image from a supercomputer simulation, the paths of blue and orange tracer particles give a comprehensive view of air motions in and around a developing severe thunderstorm. The image maps the cloud's raindrop content in greens and yellows at a horizontal plane 2.25 kilometers above the ground. Below 2.25 kilometers the surface enclosing rain and cloud drops is rendered as a solid and above as a transparent curtain.

For instance, when astrophysicists wondered whether an exotic arrangement of stars could be two colliding galaxies, they programmed a computer to calculate the behavior of two galaxies as their paths crossed. The computer generated numbers, which when transformed into images, showed whether two colliding galaxies would look like the astronomical phenomenon. Other numerical models can reveal the damage from a car crash, the action of a drug on a cancer cell, the growth of a thundercloud, or the evolution of the universe.

Our ability to create these numerical models rests on the amazing fact that mathematics can be used to represent the physical universe. No one really understands why this should be so, but much of the success of modern science and engineering is based on our ability to create an abstract mapping between the motions of matter and symbols on paper. Following this approach, scientists have gradually discovered a set of mathematical equations, generally referred to as the laws of nature, that describe the physical world. These fundamental equations are written in the language of calculus, a branch of mathematics that deals with rates of change—how one quantity varies with respect to other quantities such as time or location. Such equations are precise because calculus divides time and space into points that are infinitesimally close to each other.

A developing thunderstorm can be viewed as nature's way of physically solving a subset of the laws that govern gas dynamics, heat transfer, and the properties of water. This physical solution provides an evolving temperature, pressure, wind speed, and wind direction at each point in space. It also specifies the phase state of water—whether the water is in vapor, liquid, or solid form—and it specifies the size, amount, and kind of water droplets or ice crystals.

Imagine recording the numerical value of each of these variables at all points in space in the volume of the atmosphere containing the storm and for all moments in time during the storm's evolution. This infinite set of numbers would, in principle, constitute a digital solution

of the laws of nature. A different thunderstorm, evolving with another set of numbers, would represent a different digital solution to the same laws of nature. Thus to every universal law there are an enormous variety of possible physical solutions, each of which has a corresponding digital solution.

Although nature is continually constructing physical solutions to its laws, humans can neither experimentally nor theoretically produce completely accurate digital solutions. Our only practical approach to creating digital solutions is to use supercomputers to create a mathematically approximate solution to these equations, called a numerical solution. This process is often called simulation.

To perform a simulation, the scientist begins by choosing an appropriate subset of the fundamental equations. These equations, which are valid at every point in space and at every moment in time, are then replaced with a closely related set of equations defined only at selected points in space and selected moments in time. These "discretized" equations are programmed into the supercomputer. Instead of tackling the impossible chore of solving the laws of nature everywhere for all time, the supercomputer evaluates required quantities only at the selected points at prescribed time intervals.

Such an approach has a number of distinct advantages. First, scientists can replay a solution over and over, whereas they are able to observe most natural phenomena only once. Second, they can study an ensemble of solutions, each describing the same phenomenon but with variables of different values. For example, they might simulate many different thunderstorms in order to extract the common defining properties from the details of the solutions. Scientists can isolate a dominant subcomponent of a phenomenon and make a more finely resolved simulation of just that feature. Finally, they can determine all the values of the physical variables as these variables change through space and time.

However, the numerical solution by itself is of little use. Today's supercomputers are capable of performing one billion arithmetic operations per second, and a typical simulation runs for hours. Even a small portion of the results comprises billions of numbers. No scientist could digest the vast columns of numbers that stream from the computer programs used in simulation were they not transformed into images.

Simple arithmetic drives home the necessity for visualization techniques that display data in the form of pictures. In one second of operation, a modern supercomputer can generate one billion numbers, which if printed out in 10 columns of 50 lines apiece on each page, would require a pile of paper over 50 stories tall! As early as 1995, when supercomputers will be one thousand times faster than today, that one second of operation would produce enough printed output to rise more than 100 miles in height. For this reason, scientific visualization has become as important as supercomputers to the computational scientist.

By inserting more spatial points and shortening the time intervals, scientists can increase the accuracy of a simulation, because the discretized computation approaches the continuous coverage of space and time characteristic of the exact ideal solution. However, improving the realism of the simulation, either by adding more points and time intervals or by adding more laws of nature to the set of equations, increases computing time. It also greatly increases the amount of computation needed to transform the enlarged numerical output into visual images.

An insatiable craving for ever faster supercomputers is a direct result of the scientist's desire to produce solutions of increasingly realistic complexity during an allotted span of computer time. For this reason, scientists have enthusiastically adopted each new generation of digital electronic computers for over fifty years. During that time, computers have increased in speed by more than a billion times! And yet today's scientists are just as unsatisfied with current technology as their predecessors were on the eve of World War II.

Although at any given moment there has always been a computer that was the fastest in the world, the term "supercomputer" began to be commonly used for the fastest computers only with the introduction in 1976 of the Cray-1 supercomputer, manufactured by Cray Research. Fifteen years later, the Cray Y-MP supercomputer from the same company was 16 times faster than the Cray-1. The demand for supercomputers has now brought forth a new generation of companies, each with a "better idea" of how to make machines that are faster yet. By 1996 we can expect supercomputers to reach speeds some 500 times faster than those of only five years earlier. The pace of change is not only unrelenting, it is also drastically accelerating. As the power of supercomputers increases, so will the power of scientists to create and manipulate digital worlds at will, propelling humanity toward new levels of insight and comprehension.

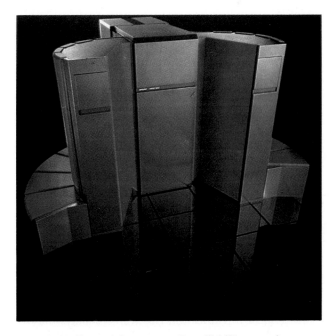

At peak speed, an eight-processor Cray Y-MP can perform two trillion arithmetic operations per second. The computer's circuit modules are located behind the vertical panels. The cushioned, benchlike arrangement around the base of the computer contains the computer's power supply, as well as some of the plumbing which circulates refrigerating coolant.

The Three Modes of Science

For nearly four centuries, science has been progressing primarily through the application of two distinct methodologies: experiment and theory. The experimental/observational mode, first exploited by Galileo in the early 1600s, uses instruments like telescopes, microscopes, and particle accelerators to search for regularities and patterns in the enormous complexity exhibited by natural phenomena. The goal of the experimental branch is to discover facts from which physical models of reality emerge. We refer to such models when we speak today of molecules, viruses, galaxies, and the age of the universe. From William Harvey's discovery of the circulation of blood to Ernest Rutherford's proof that atoms have nuclei, the experimental/ observational mode has given us fundamental insights into the world around us.

The theoretical mode, epitomized by the work of Isaac Newton in the mid-1600s, strives to encode the discovered regularities and patterns of the physical world into a set of relationships between mathematical variables. These relationships are expressed by the equations that form the laws of nature. Spectacular successes of the theoretical mode include the Euler and Navier-Stokes equations governing gas and fluid dynamics and Maxwell's equations, which completely describe the behavior of electricity, magnetism, and electromagnetic fields. Another example of a natural law is the Schrödinger equation, which embodies the tenets of quantum mechanics that describe the submicroscopic world of atoms and electrons. Finally, perhaps the grandest example is the Einstein field equations of general relativity, which relate gravity to the curvature of spacetime.

The two traditional modes of science have distinct limitations, however. For the experimenter, nature is sometimes difficult to investigate. Many of the phenomena that scientists would like to observe are too small, or too far away, or too fleeting to yield readily to scientific scrutiny. Theoreticians seeking a solution for a specific instance of a phenomenon traditionally are able to evaluate mathematically only the simplest scenarios. For instance, purely theoretical methods cannot solve exactly the equations that describe the dynamics of a thunderstorm.

The development of digital computers has transformed the pursuit of science because it has given rise to a third methodology: the computational mode. The intent of this mode is to solve numerically the theorist's mathematical models in their full complexity. A simulation that accurately mimics a complex phenomenon contains a wealth of information about that phenomenon. Variables such as temperature, pressure, humidity, and wind velocity are evaluated at thousands of points by the supercomputer as it simulates the development of a storm, for example. Such data, which far exceed anything that could be gained from launching a fleet of weather balloons, reveals intimate details of what is going on in the storm cloud. Furthermore, the computational scientist can compute the collisions of a few atoms just as easily as the collisions of enormous galaxies; the scales of space and time are simply input parameters of the computer program.

Exploring Solution Space

The mathematical laws of nature that describe the fundamental workings of natural phenomena express universal relationships between such quantities as energy, mass, momentum, temperature, pressure, and density. These equations are powerful because they govern all possible situations, but they do not tell us anything about a particular situation until we "solve" them. For example, Maxwell's equations are universal, *general* statements about electricity and magnetism, but they do not immediately tell us anything about a specific arrangement of a certain set of electric charges that we might want to investigate. If we want to know the electric field in the vicinity of these charges, we must incorporate information about the locations of the charges into Maxwell's equations and then solve them for that particular case. The data appended to the laws of physics to characterize a specific scenario include the boundary conditions that define the spatial extent of the phenomenon to be modeled and the initial conditions that give starting values for the fundamental variables.

When solving the laws of nature, a theoretical scientist traditionally attempts to find an exact solution in symbolic form; such a solution can be written down in terms of known mathematical functions, like sine, cosine, or x^2. The relative simplicity of such functions typically requires the scientist to limit the problem under investigation by making some simplifying restrictions such as requiring the geometry of the problem to have special symmetries or the system to be at rest (in equilibrium) and thus unchanging over time.

The complete collection of all solutions covering every conceivable circumstance expressed by a particular law of nature is called a "solution space." Because of the simplifying restrictions required by exact symbolic solutions, these solutions probe only a small region of solution space, and thus tell us little about how nature can behave if, for instance, the geometry of a system is highly asymmetrical or if the system is vigorously dynamic.

A technique called perturbation theory can be used to explore regions of solution space in close proximity to an exact solution. If an actual system differs only slightly (it is "perturbed") from a simpler system for which an exact solution exists, that solution may be expanded mathematically to cover a slightly wider range of circumstances. Thus, for instance, perturbation methods make it possible to investigate a system that has slightly less symmetry or is very near equilibrium. In spite of their great usefulness, these traditional techniques still leave vast areas of solution space unexplored.

By way of an example, consider water waves produced by winds blowing across the ocean. When in equilibrium, the water and air would be separated by a perfectly flat boundary. Obviously, the solution is trivial to write down: the density on each side of the interface is constant (since the two fluids do not mix or propagate waves), and the air and water are moving at a uniform constant velocity with respect to each other, or not moving at all. The familiar small-amplitude waves induced by a gentle breeze are a perturbed solution of this equilibrium state. The solution can be symbolically represented as a nearly flat interface rippled by sine waves propagating with constant velocity.

These waves represent what scientists term a Kelvin-Helmholtz instability, after two nineteenth-century physicists who studied such phenomena. By examining solutions to the equations of fluid dynamics, they discovered that small imperfections in the surface at the interface will grow to form moving waves when two gases or fluids move slowly relative to each other.

But when the gentle breeze increases its velocity to become a raging gale, the simple rolling wave develops much more complex shapes, as the ocean's surface becomes a seething tempest of breaking waves and whitecaps. Here, in a general region of solution space, there exists no exact formula to capture the complexity of the solution representing these phenomena.

From the Continuous to the Discrete

During the 1940s and 1950s, the Hungarian-American mathematician John von Neumann described a general procedure to explore those regions of solution space beyond the reach of exact solutions. To study complex phenomena, we must leave the continuous world behind and substitute in its place a discrete representation of the phenomena. Then, using a supercomputer, we can indeed solve the relevant equations of fluid flow even if the wind is blowing faster than the speed of sound!

The laws of nature govern what happens at every point in space and at every moment of time. Fortunately, to carry out our science, we do not need to compute solutions to these laws that finely. We can apply our equations at a set of points in space that are separated by distances that are small compared to the size of the objects we care to study. We can study the evolution of the system at discrete intervals of time that are short compared to the duration of the process we are exploring. This replacement of the continuous mathematics by discrete points in space and discrete intervals of time is the key that permits us to use supercomputers to explore the perennially forbidden regions of solution space.

This approach actually predates von Neumann. Consider the archetypical formula for the sine wave: sin (x), which can represent the solution to a wave on a plucked string. This formula is very convenient for symbolic computing, as when one needs to manipulate trigonometric identities in order to prove a theorem. However,

when we need to use the sine function to solve a practical problem, we evaluate the function at a discrete set of points and exhibit the solution in the familiar table of numbers found in any book on trigonometry, thereby achieving a numerical solution. Indeed one of the first uses of the early mechanical calculators hundreds of years ago was to compute trigonometry and logarithm tables from formulas.

Neither the formula "sin (x)" nor the tables of numbers by themselves immediately call to mind the beautiful regularity of this function. To "see" this regularity we graph the numbers, converting the solution from a digital to a visual form that the human brain (some half of whose nerve cells are devoted to visual processing) can easily grasp and associate with physical phenomena, such as rolling waves on the ocean's surface.

This simple methodology of first solving the mathematical laws of nature in a discrete fashion using computers, then converting the numbers to visual images so that the human mind can extract understanding, is the funda-

---

SYMBOLIC $\sin^2 (x) + \cos^2 (x) = 1$ $\sin (x + 2\pi) = \sin (x)$

NUMERIC

x	sin (x)	x	sin (x)	x	sin (x)
0.00	0.00	2.20	0.81	4.40	−0.95
0.20	0.20	2.40	0.68	4.60	−0.99
0.40	0.39	2.60	0.52	4.80	−1.00
0.60	0.56	2.80	0.33	5.00	−0.96
0.80	0.72	3.00	0.14	5.20	−0.88
1.00	0.84	3.20	−0.06	5.40	−0.77
1.20	0.93	3.40	−0.26	5.60	−0.63
1.40	0.99	3.60	−0.44	5.80	−0.46
1.60	1.00	3.80	−0.61	6.00	−0.28
1.80	0.97	4.00	−0.76	6.20	−0.08
2.00	0.91	4.20	−0.87	6.40	0.12

VISUAL

The simple function, sin (x), can appear as a term in a symbolic statement (top), or it can be evaluated at discrete points to produce a numerical solution (left), which can be converted to graphical form (right).

mental approach that we shall see being used again and again in this book. Let us now look at a more complex version of our idealized wave example, moving from exact to perturbed to complex regions of solution space. In doing so, we will see how a visualized supercomputer simulation can give us unexpected new insights into the solutions of the laws of nature.

Gentle Waves to Whitecaps: An Idealized Example

During the 1980s, a number of physicists and astronomers became interested in the phenomenon of supersonic jets, but not the kind that land at airports. To these scientists, a supersonic jet is a stream of fluid that squirts into a second fluid at a speed greater than the speed of sound. In this context, "fluid" refers to either liquids or gases, since they both obey the same basic equations. For the astronomers, the goal was to understand how enormously powerful jets of matter, which are ejected at supersonic speeds from the centers of certain galaxies, can travel relatively undisturbed for millions of light years through the gaseous medium first within and then surrounding the galaxy. We return to the simulation of the entire jet system in Chapter 3.

Instabilities that disrupt a supersonic jet begin at the interface between the supersonic fluid and the medium through which it flows. If the flow is unstable, tiny ripples on the interface grow with time into huge, crashing waves. To study such phenomena, Paul Woodward at the University of Minnesota and his colleagues chose to simulate two fluids moving past each other at extremely high speeds. To see what happens at the interface, Woodward and his colleagues developed a supercomputer simulation equivalent to a supersonic wind blowing across an imaginary ocean.

A wind blowing on small ripples exerts a force causing the pressure on their windward side to exceed on average the pressure on their lee side, and this pressure difference brings about a transfer of energy from wind to water. The energy and pressure generate alternating regions of compression and decompression in the water that create visible waves on the surface. A faster wind imparts more energy to the water, whereas a denser liquid would show greater resistance to the wind's force. The laws of nature that Woodward chose to describe this behavior of fluid flow are the Euler equations of inviscid gas dynamics, formulated by the eighteenth-century Swiss mathematician Leonhard Euler. They consist of a set of equations that relate the fundamental physical variables of the density, pressure, internal energy, and velocity of a fluid. To program a supercomputer, Woodward used a discretion technique called finite differencing, which we explain in Chapter 3, to transform Euler's equations from a statement in calculus into a set of algebraic equations that could be evaluated at discrete points.

In the real world, wind blowing across an ocean or a jet boring through a medium occurs in three-dimensional space. But the supercomputers available to Woodward and his colleagues in the mid-1980s were not powerful enough to perform a full three-dimensional simulation, and so the problem was simplified to two dimensions. Although the resulting scenario is an idealization, it still illustrates the main features of the fluid flow.

To set up the simulation, imagine a vertical, rectangular slice through the interface between two fluids. The upper half of the rectangle is occupied by a low-density fluid, while the bottom half is covered by a higher-density fluid. Woodward needed to transform this two-dimensional slice into a "computational space" composed of a set of points at which to evaluate the equations. To that end, he covered the rectangle with a grid measuring 720 points vertically by 360 points horizontally. The density of fluid in the lower half of the grid was chosen to be ten times that in the upper half, so the scenario resembles the water/air interface on the surface of the ocean. To avoid the grid having walls, Woodward allowed the waves running off the righthand side of the grid to reappear on the left. Without this so-called periodic boundary condition, the vertical walls of the grid would have caused the simulation to resemble turbulence in a rectangular aquarium rather than waves in an open expanse of ocean.

In the real ocean, waves start up when the frictional drag of a breeze on a calm sea creates ripples. Woodward's waves do not originate in this way, however. The fluids in his simulations are nonviscous, meaning that they slip past each other without any drag or shear; hence the boundary between the two fluids is called a "slip surface." The scientist must therefore help matters along by introducing a small-amplitude wave that perturbs the interface between the two fluids at the onset of the simulation.

At each time interval during the simulation, the supercomputer solves the approximated Euler's equations for the physical variables at all $720 \times 360 = 259,200$ grid points before moving ahead to the next interval. As the supercomputer generates solutions to the approximated Euler equations, the results appear on the widescreen color monitor of a workstation, where Woodward and his colleagues can watch the growing waves as one fluid races past the other. As the upper fluid continues to transfer energy to the lower fluid during the simulation, the smaller waves break and give way to larger ones. In each grid zone, color represents density, with the densest fluid shown in green and the least dense fluid in light blue. Intermediate colors represent the varying range of density as the two fluids mix.

The images on the facing page display four successive time intervals of Woodward's simulation of one fluid moving past the other at the speed of sound. The tiny initial perturbation grows in amplitude as soon as the evolution begins, displaying the characteristic Kelvin-Helmholtz instability. However, because Woodward can solve the full equations with the supercomputer, the shape and amplitude of the sine waves soon become much more complicated.

Ordinarily, a small disturbance in a fluid propagates through the fluid at a characteristic speed called the speed of sound. But a large disturbance can produce an abrupt jump or discontinuity in the fluid's density and pressure that moves faster than sound. An explosion, an earthquake, and an aircraft traveling faster than sound are familiar sources of such supersonic disturbances, which are called shock waves. The strength of a shock wave is defined by the Mach number, the ratio of its velocity to the undisturbed sound speed, named after the Austrian physicist Ernst Mach, who studied supersonic motion in the late 1800s.

Many weak shock waves appear as orange creases in the lower fluid during the earliest stages of Woodward's simulation. These shocks are associated with the development of the ripples of the growing Kelvin-Helmholtz instabilities. In the second image on the facing page, the shocks have become more pronounced in the lower fluid while weaker shocks have appeared in the upper fluid. Aquamarine areas in the upper fluid are vortices spinning so fast that they have become regions of low density, much like those at the centers of tornados.

In the third image, numerous intersecting shock waves are propagating throughout the denser fluid, which has begun to mix with the less dense fluid above. The fourth image, which is dominated by one enormous breaking wave, demonstrates that the shocks eventually organize themselves into a few very strong shock waves oriented at a specific angle. This angle, as well as the propagation speed of the shock disturbances, was later derived from a purely theoretical treatment of this problem by Andrew Majda and his coworkers at Princeton University. The agreement of angle and speed with theoretical predications serves as an important confirmation that the simulation has been done correctly. The waves certainly turn the perfectly flat interface into a ragged one, but they do not destroy the basic separation of the two fluids, and in this way conform to the cosmic jet observations.

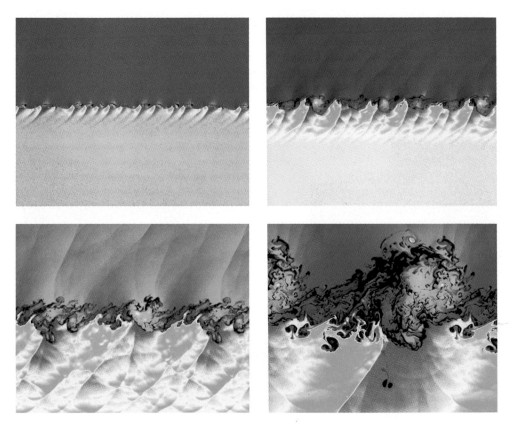

As the upper fluid flows past the lower fluid at the speed of sound, the two fluids begin to mix and Kelvin-Helmholtz instabilities grow into graceful arching waves. Meanwhile numerous shock waves in the lower fluid organize themselves into strong shocks oriented at an angle diagonal to the fluid interface. Color indicates density; blue is the least dense and green is the most dense.

The large crashing waves bear an uncanny resemblance to the artist's image on the Frontispiece. It is reassuring to see the supercomputer pull such familiar shapes out of the mathematical expressions of the laws of nature. But can we go further? Although illustrative of many of the principles of computational science, Woodward's slip-surface simulation was an idealized problem in two spatial dimensions rather than three. How far can we push this notion of deriving digital realities that display behavior very similar to the behavior of the physical realities of our experience?

The Severe Storm: Can We Simulate Reality?

The tens of thousands of thunderstorms that occur somewhere on the Earth during a single day are critical subcomponents of the atmospheric system for distributing heat, water, and electrical charge. Countless observations have revealed that the most powerful storms have recurring features.

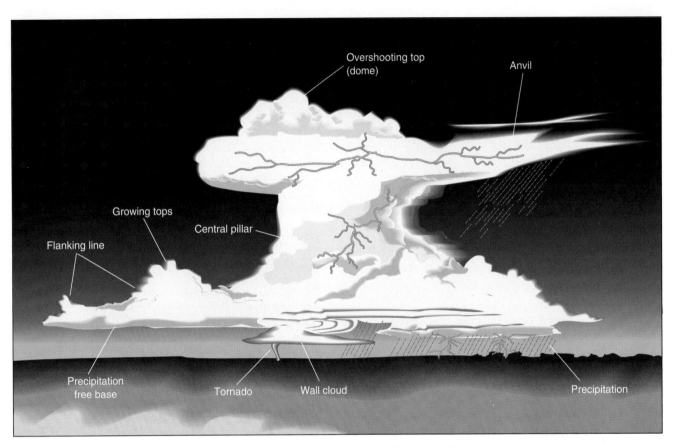

Overshooting top (dome)

Anvil

Growing tops

Central pillar

Flanking line

Precipitation free base

Tornado

Wall cloud

Precipitation

Many decades of observation have established the recurring features of a Great Plains thunderstorm.

A severe thunderstorm is a self-contained, coherent phenomenon in the atmosphere that moves at high speed while maintaining its basic structure. A powerful updraft of warm, humid air rising from near the ground creates the central pillar of the storm that dominates the illustration on this page. The cloud top forms where the temperature of the cooling, rising air reaches equilibrium with the environmental temperature. The air then begins to move sideways, forming a characteristic anvil top that is often capped by an overshooting dome located directly over the central pillar of the storm. Extending in the opposite direction from the anvil, near the base of the thunderstorm, is a flanking line of clouds created by an outflow of cold air from inside the storm. Numerous small-scale features accompany the storm, ranging from the cauliflower-shaped turrets on the cloud to occasional tornados and downbursts.

Having identified the external features of a storm cloud, scientists developed a variety of instruments to probe its internal structure. For instance, when 16 tornados struck western and central Oklahoma during the afternoon and evening hours of May 20, 1977, four Doppler radars, 45 surface stations, four instrumented balloon sites, two instrumented aircraft, and several mobile tornado chase teams observed the details of the storm system. Modern Doppler radars are able to measure not only the amount of rain in a storm, but also the local motion of the rain toward or away from the radar site.

From the views of the storm provided by these various instruments, researchers Peter Ray, Robert Wilhelmson, and Kenneth Johnson, working with the National Severe Storms Laboratory in Norman, Oklahoma, were able to construct a detailed digital map of the rain intensity and wind velocity at six different elevations within the storm. Using a novel scientific visualization technique a decade ago, these researchers glued differently colored arrows on a grid of wires to represent the horizontal wind velocity and local water density at a single elevation. They then placed the grid for each elevation, supported by strips of Plexiglas, at the appropriate height above the ground level.

The combined observations of the rain content and wind velocity of severe storms reveal some coherent internal structures. For instance,

on the mid-levels one can see a finger or "hook echo" forming in the rain field (red). Repeated observations with conventional non-Doppler radar had led to a rule of thumb that says there is a high probability of a tornado forming near such a hook. With the velocity information from the Doppler radar, we can see why this might be, since the rain falls in an area where the wind is forming a rotating vortex, although the vortex is much larger than any tornado that might ultimately form. We can also see a large rain-free (dark blue) region on the lowest level, where strong winds change direction abruptly as the rain begins. This abrupt change in wind direction is termed a "gust front," the boundary between the falling cold air and the ambient warm air beneath a storm. The computational challenge is to see if a cloud with all these external and internal features naturally arises in a supercomputer simulation built on the basic laws of nature. Accurate simulations would give scien-

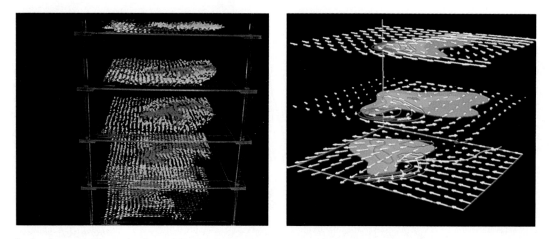

Left: *Arrows indicate the flow of air relative to the storm's movement in this wire-and-plastic display of a tornadic storm that struck on May 20, 1977. The length of arrow indicates wind speed, and the color the intensity of rainfall (including ice and hail), from rain-free (blue) to heaviest (red).* Right: *In these cross sections through a thunderstorm from a simulation performed in the mid-1980s, white arrows depict the horizontal wind flow; the red and green contours outline the regions of updraft and downdraft, respectively; and the yellow line marks the storm's gust front—the boundary between the cold air flowing out of the storm and the warm air in front of it. The blue regions mark areas of rainfall.*

tists a tool for discovering and examining the underlying mechanism that produces these structures.

The study of severe weather using three-dimensional storm models was pioneered over twenty years ago by several scientists, including Robert Wilhelmson of the University of Illinois at Urbana-Champaign and Joseph Klemp of the National Center for Atmospheric Research (NCAR). The evolution of a storm can be simulated by solving the equations for compressible gas flow, just as Woodward did above, along with the equations that give the behavior of the three phase states of water (vapor, liquid, and solid ice particles) at different temperatures and pressure. The supercomputer solves these equations in discretized form, computing the values of water and gas variables such as air pressure, temperature, and wind velocity. A three-dimensional grid divides the three-dimensional rectangular volume that encompasses the whole storm into small boxes called volume elements, and the variables are determined at discrete time intervals in each discrete box.

Scientific computer graphics had progressed enough by the mid-1980s that, instead of visualizing the simulated data by constructing Plexiglas models, Klemp could plot wind flow and water density at various heights on the same screen using graphics packages his group had developed at NCAR. When Klemp compared his simulations with observations, he could see the development on the second level of an air vortex that creates a hook shape in the region of condensed water (blue shading, see previous page). On the lower level he could clearly see a gust front. Both of these simulated features appear just as seen in the Doppler radar observations!

Simulations showing the development of the storm through time matched observations as well. In 1980, Wilhelmson and Klemp modeled a well-observed storm system that had devel-

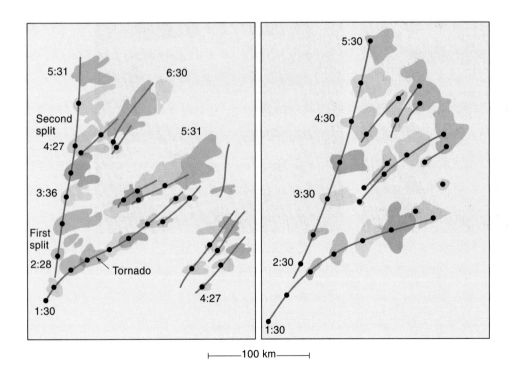

The development of a storm on April 3, 1964, as observed by radar (left) and as simulated with a supercomputer (right). The areas of differing color represent individual storm cells, which in the map on the left correspond to regions of strong radar reflectivities and in the map on the right to heavy rainfall 0.4 kilometer above the ground. Points of greatest reflectivity or rainfall are connected by solid lines, to indicate the storm tracks of the splitting storm system. A given color represents the most intense portion of the storm at a particular moment in time.

oped over Texas on the afternoon of April 3, 1964. It unleashed strong winds, heavy rain, and hail, and spawned a tornado that killed 7 people and injured 111 others, causing over $15 million in property damages.

Non-Doppler radar followed the development of the storm by revealing regions of larger raindrops, ice, and hail in a horizontal plane near the ground. It showed the young storm quickly elongating and splitting into two storms by 2:28 p.m. Eventually additional storms developed, causing a long line of strong storm cells (dark shaded regions) to appear by 5:31 p.m. The simulation storm path develops in a remarkably similar fashion, showing that such models can capture the major features of the behavior of a complex storm system.

Before these early simulations were performed, it was not obvious that such simple models would match observed storm behavior, since many of the physical processes that scientists know about were not included in the model. For example, the effects of surface friction were not considered, the influence of the earth's rotation was ignored, turbulence at scales smaller than the grid boxes could not be resolved, and the development of water and ice in the model were only crudely represented. However, it is apparent that the overall evolution of the storm can be successfully simulated without considering these details.

We will encounter this remarkable hierarchical property of nature throughout the book. We could simulate the water molecules in the thunderstorm by themselves perfectly well, or we could study the deformation of a single raindrop as it accelerates through the air during its fall, or we could study the entire Earth's atmosphere with an approximation for the tens of thousands of thunderstorms taking place during the day. Nature seems to be constructed like Russian dolls nesting within each other. One can pick out a doll of any dimension and appreciate it for itself. Unlike the dolls, however, one level of nature couples to the next layers above and

below it in scale, and these interactions are described by the laws of nature. If natural phenomena did not have this hierarchical character, the practice of science itself would be nearly impossible.

The Importance of Speed and Memory

Level crosscuts like Klemp's mid-1980s visualization on page 11 provide insight into the velocities within each two-dimensional level, but they do not display the flow out of the plane. Similarly, when viewing crosscuts through a storm, we can understand where the three-dimensional surface of the cloud intersects each layer (the

In 1987, advances in software allowed a visualization team led by Stefen Fangmeier at the National Center for Supercomputing Applications to produce a three-dimensional animation of a storm's evolution from its early life to maturity. This image from the video shows the surface inside of which water is in the form of large raindrops.

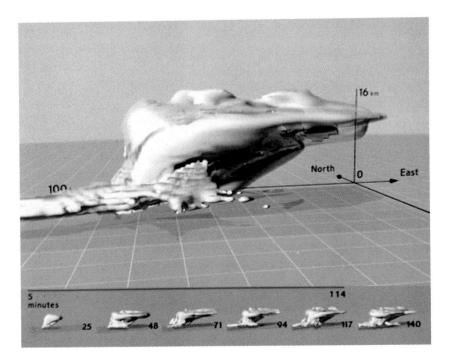

Information designer Edward Tufte collaborated with graphics specialists at the National Center for Supercomputing Applications to create a presentation format that shows the simulated cloud volume in space and time. The slightly darkened surface indicates the base of the computational grid (100 by 54 kilometers); the grid extends 16 kilometers vertically, and vertical height is exaggerated by a factor of two. The grid moves with the storm as the storm travels. The surface has been further darkened wherever there is cloud, ice, or rain directly above. The views below show six successive stages in a developing storm; the end of the red line locates the top image in the sequence.

blue areas in Klemp's image), but we do not get a geometrical picture of the shape of the cloud's surface or the contours of water density within it. By the late 1980s, Wilhelmson's team was able to produce realistic three-dimensional images of the storm that allowed an easy comparison with actual storms. From the outside, the simulated storm develops a shape almost identical to the schematic diagram on page 10. One clearly sees the flanking line, central pillar, overshooting dome, and anvil top. This structure is remarkably stable over the two hours after it first develops from a small cumulus cloud.

Since the early 1970s, when pen plotters slowly drew black-and-white contour plots to represent water density or wind velocity, our ability to create useful visualizations has undergone a revolution, even though the method for solving the laws of nature describing a thunderstorm has changed very little. The transformation has come about in part because inventive graphics experts have created graphics software able to produce fully rendered surfaces replete with light sources and shadows. But their efforts could not have borne fruit if the speed and memory of supercomputers themselves had not increased greatly. Even now a graphics supercomputer may take several minutes to compute a single image from the already computed storm data. As late as 1987, an animation of an evolving storm required eight hours or more of graphics computation, longer than it took the supercomputer to solve the laws of nature in the first place.

The simultaneous development in the late 1980s of both new capabilities in computer graphics and very large memory supercomputers prompted Robert Wilhelmson to simulate the evolution of an individual severe thunderstorm in detail. Led by Wilhelmson and visualization specialist Matthew Arrott, this effort teamed scientific researchers, computer graphics specialists, animation and choreography experts, and video post-production staff. The resulting video animation became an international sensation, awakening many people's imaginations about what could be accomplished with supercomputers.

To begin the simulation, Wilhelmson and his colleagues filled the model volume with data describing wind, temperature, and humidity obtained from a balloon sounding made on the day of the storm. Warm moist air near the surface coming from the south is overlain by drier, rapidly ascending air from the west. These initial conditions are perturbed at the center of the volume to create a tiny cumulus cloud, which then evolves into a severe thunderstorm. The large-memory Cray-2 supercomputer can update the variables of the model twice as fast as the natural storm can develop.

To visualize the activity inside a storm having the characteristic outside shape shown on the facing page, the surface of the cloud is rendered as a transparent three-dimensional surface. The air flow inside the cloud can be seen by using the computed velocity field of the air to track the positions of weightless tracer particles; the particles are released from a horizontal plane one kilometer off the ground. Yellow streamers trace out the trajectories of selected particles, giving a view of the major air currents within the storm. These visual indicators enable the researchers to decode the storm's underlying mechanisms.

The simulation shows the storm growing quickly from the cumulus cloud as warm humid air condenses in the central updraft. When the upwardly moving saturated air reaches an altitude in the stratosphere where it is no longer warmer than the surrounding air, it spreads out to form the anvil top. The updraft, propelled by

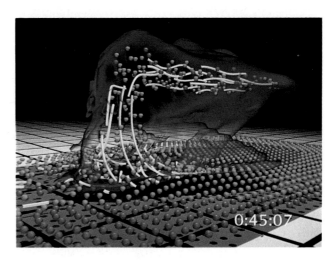

The positions of weightless particle tracers, shown as colored balls, depict air motion in a severe thunderstorm, after 15 minutes of simulation, when a small cumulus has developed (left), and at 45 minutes as the anvil of the storm begins to form (right). Before release, the tracers are spaced at 2.0-kilometer intervals on a horizontal plane 1.0 kilometer above the ground. Orange tracers are moving up, and blue down at the time of viewing. Yellow streamers trace the paths of selected particles over a nine-minute period.

its strong momentum, overshoots the anvil top, creating a dome shape above the anvil just as seen in the physical object.

When the suspended water drops become heavy enough, rain begins to fall, dragging the adjacent air downward in a strong downdraft. The downward moving air twists around the upward moving air in a pattern originally hypothesized from non-Doppler radar. The cold air in the sinking downdraft supports a flanking line of clouds near the base of the thunderstorm along the gust front. As this cold air meets the warm, moist air near the ground, it lofts the warm air upward, feeding the updraft of the central pillar and closing the circuit. This separation of updraft and downdraft is crucial to sustaining the life of the storm.

Changes in the phase states of water help to create a self-sustaining cycle. The heat released as the water vapor in the rising air condenses into droplets adds to the central pillar's buoyancy. In a reverse fashion, the falling rain cools the surrounding air by evaporation, reinforcing the cold air's downward motion. The equations describing the changing phase states of water must be correctly coupled with the equations describing air flow to compute an accurate simulation of these updraft and downdraft motions.

The rotation of air in and around the simulated storm is revealed with broad orange ribbons. An individual ribbon's rate of twist indicates the amount of local rotation, called stream-wise vorticity. This twisting is similar to that of a perfectly thrown spiraling football. The rate of rotation grows substantially as the ribbons begin to rise in the updraft or enter the downdraft. Such rotation in these long-lived storms can ultimately give rise to large tornados.

This visualized simulation has laid bare the inner dynamics of a common natural phenomenon. Although the outside of the storm was familiar to everyone, its internal mechanisms had always been masked by the opaque cloud

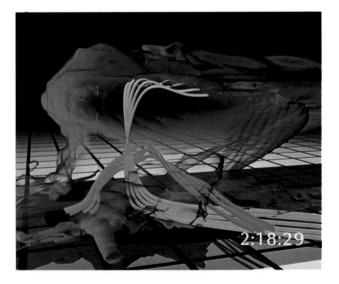

Orange ribbons represent tracers that rise through the depth of the storm in the updraft and blue ribbons represent tracers that eventually fall to the ground in the downdraft in this image tracing particle paths from approximately 75 minutes after the start of the simulation to 2 hours and 18 minutes after.

Tracer ribbons illustrate the development of streamwise vorticity during a 20-minute period. The relative magnitude of the vorticity is largest in the lower part of the storm, where the ribbon is most tightly coiled.

itself. Such storm simulations have been used to verify hypotheses for how air moves within storms, to identify what physical forces govern the direction that severe storms move, to understand why storms in hurricanes can produce tornados under conditions substantially different from those that produce tornados in the Great Plains, and to explore why tornados develop in some severe storms and not in others.

The work of Wilhelmson and his colleagues over the years illustrates how the increase in the speed and memory of supercomputers, in combination with new computer graphics techniques, have led to ever more revealing images of severe storms. The same increase in speed and memory can have an equally dramatic effect on the underlying computations.

In 1972, Wilhelmson ran his thunderstorm simulation on a Control Data Corporation 7600 supercomputer, which had a main memory holding at most 0.064 megaword, or 0.064 million words. This memory was so tiny that he had to insert explicit commands into his program to shuffle data into and out of core memory to an extended memory of 0.5 megaword, much as you would write and read to a disk. He was able to compute the dynamics of a volume of air 14 by 38 kilometers on its base and 10 kilometers high. His computational grid divided the volume into volume elements 600 meters on a side at the base and 500 meters tall, yielding approximately 30,000 volume elements. Although the computer's two-stage memory could hold a volume only barely large enough to enclose a thunderstorm, Wilhelmson was able to evolve the atmospheric condition to the point that a thunderstorm developed. He was able to compute about one hour of thunderstorm evolution in about one hour of computer time.

By 1990, the main memory of supercomputers had increased a thousand times! There was no more need to shuttle data between two memories, greatly simplifying the code writing. The larger memory and faster speed of the Cray-2 supercomputer allowed Wilhelmson to enlarge the volume modeled to 54 by 100 kilo-

meters horizontally and 16 kilometers vertically. The storm could now develop to its full extent without running into the edges of the volume until late in its evolution.

The volume zones were no finer in size than they had been in 1972, and indeed were somewhat larger (1 kilometer square horizontally by 500 meters vertically). But because of the larger total volume, there were now nearly 180,000 volume elements in this calculation. Nonetheless, to compute one hour of storm evolution now required only 30 minutes because of faster hardware and faster algorithmic techniques. One was able to compute a more complex simulation in less time, a double advantage.

The fineness of the grid determines the resolution of the simulation, its ability to resolve fine details. A finer grid leads to a more accurate simulation because it more closely approximates the continuum of space and time described by an exact solution. Thus, for all the magnificence of its images, the 1990 model of a thunderstorm did not have better resolution overall than the 1972 model had, although small-scale features such as tornados could be roughly approximated by adding nested grids. For example, Wilhelmson could have placed much finer zoning in the region of the storm where intense rotation begins to develop.

Let us now consider what will be possible on a supercomputer available in 1995, only a few years from now. Because of the national commitment embodied in the High Performance Computing and Communications initiative described next in this chapter, an unprecedented increase in computer speed is currently underway. Both the peak speed and the slower simulation speed of the 1995 supercomputer are likely to be 3000 times that of the single Cray-2 processor that Wilhelmson was using in 1990.

There are several strategies for taking advantage of this awesome computing power. By simply keeping the vertical and horizontal resolution of the grid fixed at 0.5 kilometer and 1.0 square kilometer respectively, one can increase the area of the model box's base by 3000 times,

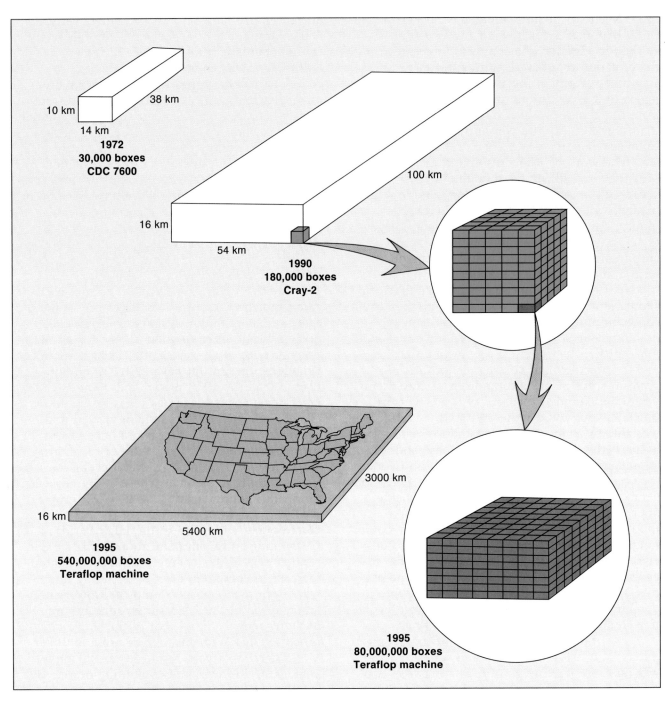

10 km
14 km
38 km

1972
30,000 boxes
CDC 7600

100 km

16 km
54 km

1990
180,000 boxes
Cray-2

3000 km

16 km
5400 km

1995
540,000,000 boxes
Teraflop machine

1995
80,000,000 boxes
Teraflop machine

*Faster computers with larger memories allowed the computational volume of a
thunderstorm simulation to enlarge considerably between 1972 and 1990. The explosion in
supercomputing speed expected to occur between 1990 and 1995 will make possible either
a huge computational volume (bottom left) or a much finer grid resolution (bottom right).*

going from 54 by 100 kilometers to 5400 by 3000 kilometers. This is a region larger than the entire continental United States! With a volume this size, meteorologists could run a severe storm forecast for the entire country and still compute faster than storms develop. To turn this dream into reality, supercomputers will have to be coupled over a high-speed national network to the next generation of Doppler radars and other observing instruments, which can provide information about local initial conditions to the models. Although creating the network will probably take us into the next century, we can seriously begin to plan for such a breakthrough.

Alternatively, modelers could put many more spatial points into the 1990-sized box, so that the grid spacing becomes much smaller. A 3000-fold increase in speed will enable Wilhelmson to cover the same total volume he covered in 1990 with volume elements that are eight times smaller in each dimension, and he will be able to decrease the time intervals between successive updates of the physical variables by eight times as well. The total number of volume elements will climb from 180,000 to over 80 million for a simulation. Storm phenomena will come sharply into focus, and the simulated cloud will begin to look very much like the real thing. Instead of nested grids, one can provide uniform coverage of the storm, depicting all the small-scale thunderstorm phenomena shown on page 10.

By allowing improvements in both scientific visualization and the underlying computations, increases in supercomputer speed and memory have a profound impact on what research scientists can perform. With each increase in supercomputer power, computational science brings the observational world of data and theoretical world of the laws of nature closer together.

The Growing Availability of Supercomputers

Supercomputers have a high capital cost, and large teams of people are required to operate

them. Thus, for three decades, from the 1950s through the 1970s, only governments could afford to purchase them, first to aid in the design of nuclear weapons, in code breaking, and in other national security applications; then later in national laboratories set up to create controlled thermonuclear fusion, to predict the weather, and to model the climate, the ocean, and the Earth's interior. During these decades, the United States government encouraged leading computer manufacturers to create ever more powerful supercomputers to meet its needs.

Supercomputers began to be employed in high-capitalization American industries, such as the petroleum and automobile companies, in the late 1970s. Engineers from these industries ran simulations of automobile performance or oil field geology. By the early 1980s, the use of supercomputers was spreading rapidly to the aerospace, energy, chemical, electronics, and pharmaceutical industries. By the end of the decade, there were hundreds of supercomputers in American corporations, and industries throughout the world had adopted the machines. Indeed, there are now far more automobile companies who own supercomputers outside of the United States than in it.

While industry was discovering supercomputers, the machines were also being acquired by a number of open scientific research centers in Europe. In the United States, however, until the mid-1980s most academic researchers had to travel to research centers in Europe or to federal laboratories to gain access to supercomputers, since very few universities could afford to purchase them. Reaching to the demand, the National Science Foundation held a national competition in 1985 that resulted in the creation of five national supercomputing centers located in San Diego, California; Pittsburgh, Pennsylvania; Princeton, New Jersey; Ithaca, New York; and Champaign-Urbana, Illinois, where the center is called the National Center for Supercomputing Applications (NCSA). In 1990, funding for all but the Princeton Center was renewed until 1995.

The mission of these centers is to give peer-reviewed researchers access to advanced computational resources, to train scientists and educators in their use, and to develop new software that will enable scientists of all fields to better attack the unanswered questions in their disciplines. Only a few years after the centers opened, the number of academic researchers in the United States using supercomputers had increased from a few hundred to more than 15,000, representing several hundred universities. To industry, the centers offer training and the chance to try out different supercomputer architectures on practical problems before making a major capital investment in equipment. In response, more than fifty corporations have joined the National Science Foundation centers as industrial partners.

The existence of the supercomputer centers has stimulated the creating of a national network, called NSFnet, linking members of the research community, not only to remote supercomputers, but to each other as well. Any desktop personal computer or scientific workstation can serve as a window into a national system of computers and users when linked to a high-speed network, either through a phone modem or by connecting with a departmental computer network. Unlike the scientist of ten years ago, who could carry out computational research only by traveling to the few national and international supercomputer centers, the scientist today "brings the supercomputer home" and appears to be using it as if it were sitting inside his or her desktop machine.

The United States government launched the NSFnet by funding the main trunks, or backbone, of the network and helping regional networks become established. Its efforts stimulated states, universities, and colleges to spend ten times the federal government's original investment in creating tributaries that now connect the desktop machines of most academic researchers to the NSFnet. Today data flows through the backbone of this network at a speed almost a thousand times faster than it flowed over the primitive link connecting the supercomputing centers in 1986. Use of the network is increasing at an astonishing rate of 10 percent compounded monthly.

The NSFnet, in turn, is part of a much larger network called the Internet, which connects the networks of other federal agencies and, increasingly, other countries. Electronic mail travels from Urbana, Illinois, to Australia as quickly and simply as from Urbana to Chicago. In Europe, South America, the Pacific rim, Canada, and Japan, governments are creating national networks resembling that in the United States to connect research universities and corporate laboratories with national supercomputer centers. By 1990, nearly fifty universities around the world had acquired supercomputers.

A desktop computer can accomplish a great deal on its own, outside the network; a powerful workstation of today has the capability of a supercomputer center of the mid-1970s. The CDC 7600 that calculated Wilhelmson's first thunderstorm had a total extended memory that is half the memory of the Macintosh Powerbook notebook computer that one of the authors used to write this book. The three-dimensional interactive graphics available on a graphics workstation for under $10,000 today can routinely create scientific visualizations that would have been far beyond the capabilities of the most advanced supercomputer center of even ten years ago.

This book records the fruits of the newly acquired access to powerful computers by providing a glimpse into the scientific and engineering research enabled by these machines. Supercomputers are helping scientists determine the structures of molecules, atoms, and atomic nuclei, and the results will give us the ability to create new materials with remarkable properties. These machines are making possible the new field of computational biology, which will examine issues ranging from the structure of macromolecules to the detailed functioning of the human body. Supercomputers enable engineers to design products that are safer, less polluting, and more efficient. They are an essential part of

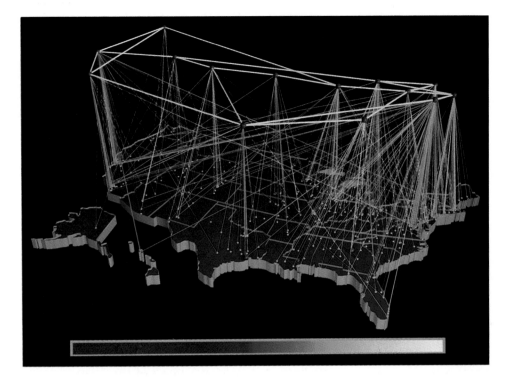

Links in the NSFnet are color coded according to the palette at the bottom to indicate the volume of inbound traffic flowing in the month of September 1991. The volume of traffic ranged from zero bytes (purple) to 100 billion bytes (white).

the continuing quest to better understand the weather, the climate, and the phenomenon of global change. Indeed, they are advancing our understanding of the nature of the universe itself.

A Look at the Future

In the late 1980s, a wide spectrum of national leaders in the United States began to recognize the strategic importance of information technologies. Legislative efforts led by Senator Al Gore and Congressman George Brown found common cause with an interagency study carried out by the White House's Office of Science and Technology Policy, and the result has been the adoption of a long-range plan called the High Performance Computing and Communications (HPCC) Initiative, which is creating a new national infrastructure to support computation and computer communications.

The HPCC program, which was initiated in October 1990, is being implemented by a large number of federal agencies that are targeting resources to activities that will advance HPCC goals. Its plans call for doubling the annual expenditures of the participating agencies from roughly $500 million annually in fiscal year 1991 to more than $1 billion annually by fiscal year 1996. The coordination of planning among the agencies breaks new ground in its approach to the management of technology by the federal government.

One goal of the HPCC program is to increase useful computing capability a thousand-fold by 1996. Program planners expect to rapidly accelerate the annual increase in speed provided by high-performance computers by effecting a radical shift in supercomputer architectures to scalable massively parallel machines. This architectural shift, described in detail in Chapter 2, will enable researchers to handle the computational demands of a set of science and engineering problems collectively called the "grand challenges." These are fundamental problems that are important to satisfying societal needs, maintaining industrial competitiveness, and continuing basic research and whose solutions require a significant increase in computational capability. They include such problems as predicting climate change, designing effective drugs, and determining the origin of structure in the universe. In addition to encouraging hardware innovation, the HPCC program supports the development of new software to enable the massively parallel supercomputers to tackle grand challenge problems and to cope with the enormous digital output of such machines.

To handle the complexity of these research problems, teams will be formed whose members come from a variety of disciplines and from multiple institutions, separated geographically from each other and the computational resources. Team members will communicate with supercomputers and with each other through a National Research and Education Network (NREN), to be created by integrating the NSFnet with the networks of other federal agencies and extending the resulting network to a larger community of sites engaged in research and education. The HPCC program envisions the NREN backbone achieving a speed of transmission a hundred times faster than today's networks, with feeders ultimately connecting the computers of millions of users.

Large-scale initiatives similar to the HPCC are now being formulated in Europe and Japan, while smaller versions are being set up in countries like Malaysia, Singapore, Taiwan, Korea, Australia, Canada, and Mexico. Many of these countries are also experimenting with advanced architectures. In Japan, the companies Fujitsu, NEC, and Hitachi manufacture computers similar in architecture to those from Cray Research, while massively parallel or neural-net based architectures are being introduced by these and other Japanese firms. In Europe, companies such as Great Britain's Meiko Scientific and Parsys, Germany's Parsytec, and France's Telmat Informatique are working in various consortiums supported by the European Economic Community to bring forth new massively parallel supercomputers.

With new technologies will come not only increased speed but improved accessibility. Soon we will even break the bonds of home and office as digital cellular telephonic technologies enable users to "jack in" to the national information network from any place at any time. Alan Kay at Apple Computer sees the next stage beyond personal computers as what he terms "intimate computing." Before the end of the century, notebook computers will have gradually acquired a color screen, pen, and voice input and output, and so much memory that it will appear essentially infinite. Combining telephone, fax, and video, these devices will become the scientist's constant companion and research assistant. Scientists will have instantly available any papers or research resources that they have been working on, and they will have a communications window to anyone or any information that they require, ultimately from anywhere in the world.

More speculatively, a number of future thinkers, such as Vint Cerf and Robert Kahn of the Corporation for National Research Initiatives, have developed the notion of software knowledge robots, termed "knowbots." These software systems will gradually customize themselves to an individual's particular style of working: they will roam the network looking for information that the individual requires and then process and package it into the format that they know he or she prefers. Although this vision dates back at least to Vannevar Bush's World

Net of the late 1940s and more recently has been dramatized by Apple Computer in its Knowledge Navigator video of the 1980s, it will probably not be realized until early in the next century.

Finally, it seems inevitable that this digital world of knowledge must transform the process of education, so that we can produce a scientifically and technologically literate society. Many school children today are totally at home in an interactive computer-based video world. The challenge we face is to create interactive software that is as compelling as video games, but that teaches science, history, music, mathematics, or social values.

The National Science Foundation supercomputer centers are engaged in a wide range of education-related activities such as producing just such software. In the future, schools whose personal computers are connected over the network to supercomputers will have "virtual" physics, chemistry, and biology laboratories created out of software in their classrooms. For instance, by running modified versions of Wilhelmson's research program students will be able to explore the development of severe storms.

In only a few years, supercomputers have made a notable contribution to the advancement of our scientific understanding. But in this book we are as concerned with what will be as we are with what is. Every question answered makes us more aware of those that remain. As we look at some of the problems now being investigated with the help of supercomputers, we will see not only where these machines have taken us so far, but where they can take us in the future.

The Evolution of Supercomputers

2

History reveals a clear pattern in the development of computers. Processing power increases rapidly after the introduction of a fundamentally new technology. The rate of growth eventually slows as the technology is exploited to its full potential. Meanwhile, new technologies are incubating, and one ultimately supersedes the others to become the new dominant technology as the cycle is repeated. Under the right conditions, the shift to a new technology can be accelerated, resulting in speed increases of a hundred- to a thousand-fold in only a few years. Thus the first all-electronic computer built with vacuum tubes was roughly a thousand times faster

The CM-5 is a scalable, massively parallel supercomputer from Thinking Machines Corporation. Each cabinet in the "lightning bolt" of four cabinets can hold up to 128 processing nodes with their local memory.

than its electromechanical predecessor, and the design of the first fully transistorized computer called for speeds nearly a hundred times greater than those of existing vacuum-tube computers. A new technology usually allows computer components to be reorganized into a new architecture, often demanding new kinds of software.

For many years, these breakthrough computers were the carefully nourished offspring of government encouragement. The needs of the war effort led to the creation of vacuum-tube computers in 1946, and transistorized computers were later developed to meet the demands of the nuclear weapons program. The U.S. government's role in the development of computer technology is a recurring theme in the history of supercomputing, and this motif is once again coming into play.

As this book is being published, we again find ourselves in the midst of another rapid increase in supercomputing power. The foundations for this latest jump were laid in the 1970s with the creation of the microprocessor, a single silicon chip that contains an entire computer. Today manufacturers are joining together large numbers of these microprocessors to create supercomputers that embody a new architecture called "scalable massive parallelism." This new design is expected to increase performance a thousand-fold within the next three or four years. Once again we find a government program (the HPCC) fueling the speedup.

Conceptual Foundations

The use of mechanical devices to speed up arithmetical calculations dates back to the invention of the abacus in ancient China. Still in use today in the Far East, this device assisted Europeans in addition and subtraction until well into the eighteenth century. In 1642, Blaise Pascal created the first true machine that could perform addition; he put digits around successive wheels, then turned the wheels to accurately sum the digits of the individual numbers. Thirty years later, the German mathematician Gottfried Wilhelm Leibniz took up the study of astronomy and was soon distressed at the tedium of computing astronomical tables. "For it is unworthy of excellent men," he wrote, "to lose hours like slaves in the labor of calculation which could safely be relegated to anyone else if machines were used." In 1672, Leibniz built a computing machine that introduced a special gear that performed as a mechanical multiplier. Although this machine could perform all four arithmetic functions, thus becoming the precursor of modern mechanical calculators, it was never dependable because of cumulative inaccuracies in the gear positions.

A portion of Charles Babbage's unrealized Difference Engine.

The outline of a general-purpose computing machine was first sketched by the eccentric English inventor Charles Babbage. In 1822, Babbage produced a mechanical calculator, the Difference Engine, which not only performed addition and multiplication but could actually solve polynomial equations. Like earlier machines, Babbage's contraption consisted of intermeshing gears turned by a crank. The British government, setting a trend that continues to this day, repeatedly gave Babbage grants of money to make a larger and faster version of his prototype. However, the component technology of that day could not achieve the precision needed to successfully build a larger working version. Furthermore, the machine suffered from limitations common to all mechanical calculating devices of that time: it could perform only specific functions, such as addition or multiplication, and the numbers had to be entered by hand.

While struggling with the Difference Engine, Babbage began to conceive of a more versatile machine that would be capable of performing any sequence of arithmetic operations. One would feed patterns of operations into this "universal calculating machine," and the machine would follow these patterns in performing its calculations. Babbage's inspiration came from the French, who, some decades earlier, had learned how to create looms that could weave any specified pattern. Indeed, Babbage realized that the punched cards invented by Joseph Jacquard in 1801 to "program" the French looms could be used to encode a sequence of operations and the numbers on which the operations were to be performed. Although he never succeeded in building one, Babbage's "Analytical Machine" was the first expression of an all-purpose, programmable computer. Its arrangement of gears and levers incorporated five basic components (input, control, processor, storage, and output) that are common to computers today.

Inventors inspired by Babbage's intellectual breakthrough made little progress as long as they tried to implement his principles within the mechanical world of gears and levers. The breakthrough finally came in the late 1930s, when it became clear that electrical rather than mechanical components could provide the underlying mechanism for computation. This strange interplay between abstract models and the nitty-gritty realities of manufacturing has shaped computing devices throughout their history.

While Babbage pondered the structure of his computing device, he was fortunate to inspire a brilliant young countess, Lady Ada Lovelace, to think about the abstract structure of the instruction stream that would control the Analytical Engine. She realized that, out of the universe of all possible instruction sequences, a few fundamental patterns emerged. The most critical of these were the loop and the conditional branch, the intellectual antecedents of the pillars of modern programming, the DO loop and the IF statement. A DO loop, discussed later in this chapter, is a very practical way of repeating a set of instructions; an IF statement is a rule by which a computer selects one of several choices. By recognizing the logical structure of programming, Lovelace laid the foundation of modern software.

Independently of this practical work, the British mathematician George Boole was studying the abstract foundations of logic itself. Formal logic explains how to distinguish valid arguments from invalid ones, by setting forth principles for drawing a conclusion from premises. An example of a valid argument is: All men are mortal; Socrates is a man; therefore Socrates is mortal. In the late 1600s, Leibniz had conjectured that logic, which had been carried out in the form of language since the time of Aristotle, could be reduced to precise, mathematical statements. In pursuit of this goal, Boole demonstrated that logical propositions could be represented with symbols like x and y. But unlike algebraic symbols, which might take on any numerical value, a logical proposition can have only two values: TRUE and FALSE. Nevertheless, Boole showed that his symbols obey laws of combination, just as numbers can be added and subtracted. For example, the AND function can join two propositions (x and y) to create a third (z): if x AND y then z. If x and y are true, then z must also be true, as in the example involving Socrates' mortality. Using the logical AND and two other operations, OR and NOT, Boole was able to construct an algebra for analyzing the validity of logical propositions.

Boolean algebra set the stage for modern computer design because numbers can be represented by a two-value system, just as logical propositions are either TRUE or FALSE. Our familiar number system is based on powers of ten (using the digits 0, 1, 2, 3, 4, 5, 6, 7, 8, and 9), but any quantity can be represented equally well by binary numbers, which use the digits 1 and 0 to stand for powers of two. For instance, in base ten the number 347 means $(3 \times 10^2) + (4 \times 10) + (7 \times 1)$, whereas in base two the same quantity is represented by the binary number 101011011, namely $(1 \times 2^8) + (1 \times 2^6) + (1 \times 2^4) + (1 \times 2^3) + 2 + 1$.

In 1867, the American logician Charles Sanders Peirce noticed that Boole's two-valued logic lent itself readily to a description in terms of electrical switching circuits. Just as a proposition is either TRUE or FALSE, a switch is either closed or open and the resulting electric current is either ON or OFF. With wires and switches, you can build simple circuits that would pass or stop currents according to the rules of Boolean algebra.

These circuits consist of elements called logic gates that perform the functions of the logical operations. The AND logic gate, for example, consists simply of two switches connected in series. Each switch represents a value of 1 or 0, TRUE or FALSE, depending on whether it is closed or open. For a current to flow through the circuit containing the AND logic gate, thereby producing an ON or TRUE value of 1, both switches must be closed. Thus, the AND operation applied to two TRUE values yields a TRUE value, as expected.

Peirce's discovery lay dormant until the mid-1930s when Claude Shannon at MIT, John Atanasoff at Iowa State College, and the German engineer Konrad Zuse independently demonstrated that binary numbers and Boolean algebra were both ideally suited for computer design. That is, an appropriate electric circuit of switches could simulate both arithmetic and logic.

In 1937, George Stibitz, a mathematician working for Bell Telephone Laboratories, took the final important step. While working at his kitchen table one evening, Stibitz hooked together some batteries, lights, and wires according to the principles of Boolean logic. By connecting several logic gates in a particular arrangement, he created a circuit that would add binary numbers. This circuitry, called a binary adder, is the basic building block of all modern computers. Similar clever arrangements of logic gates designed by Stibitz and Samuel Williams, a Bell switching engineer, could subtract, multiply, and divide binary numbers. All the ideas necessary to create a working computer were in place; it remained only to give them physical reality. However, two generations of full-scale computers would be built before the simplicity of binary arithmetic compelled its adoption by all digital computers.

The Birth of the Digital Computer

In America, the construction of large computers progressed in two giant leaps during World War II. During the late 1930s, a young Harvard mathematician, Howard Aiken, sought to construct a programmable computer that used adding wheels controlled by electrical impulses. The current was directed among the wheels by switches in the form of electromechanical relays built by the telephone company. Although the bulky relays were halfway from the mechanical world of cogs and gears to the electrical world of tubes and transistors, Aiken's machine demonstrated that the notion of a general-purpose analytical engine proposed by Babbage eighty years earlier could in fact be realized.

The project started with a generous grant from IBM, a company then renowned for its adding machines, tabulators, and typewriters. Aiken's task achieved sudden importance when the advent of World War II brought with it the urgent need for mathematical tables for calculating the trajectories of artillery shells. Development proceeded with few snags, and the computer, called the Mark I, went into operation in 1943. Operators had to set 420 dials by hand to program the Mark I, and when the computer was running, its 3304 relays clattered incessantly. In spite of its size and noise, the Mark I was an enormous success. It could handle numbers 23 digits long, adding or subtracting them in only 0.3 second, or multiplying them in 3 seconds. Following wartime duty producing ballistics tables, the Mark I continued to find service at Harvard for fifteen years.

Inside the Mark I, electromechanical relays routed electricity through a labyrinth of circuitry. However, electromechanical switches are bulky and slow. A much faster switch is the triode vacuum tube invented in 1906 by Lee De Forest. In 1943, John W. Mauchly and J. Presper Eckert were awarded an Army contract to build the first fully electronic computer, replacing relays with vacuum tubes.

Mauchly and Eckert were members of the Moore School of Electrical Engineering at the University of Pennsylvania, which had been approached to build the computer as the Army began to fall seriously behind in computing the endless number of trajectories possible with the war effort's expanded arsenal. Herman Goldstine, an army lieutenant with the Ballistic Research Laboratory of the Aberdeen Proving Ground, had convinced the Army to finance a vacuum-tube digital computer that would be orders of magnitude faster than the Mark I.

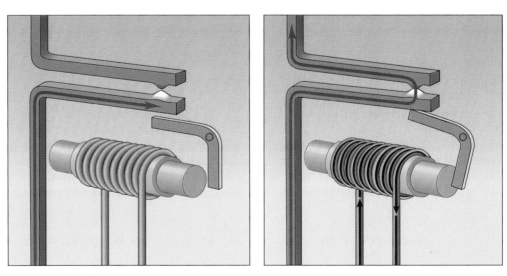

No current on coil   With current on coil

A relay is an electromechanical switch. When a current flows through the coil, the resulting magnetic field attracts the iron pivot, closing the circuit.

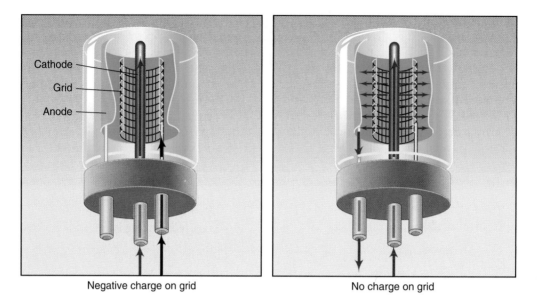

| Negative charge on grid | No charge on grid |

A triode vacuum tube is an electronic switch. The flow of electrons from the cathode to the anode is controlled by the electric charge on the grid. If the charge on the grid is negative, it repels electrons from the cathode and the current fails to flow. If there is no charge on the grid, electrons move through the grid to the positively charged anode, thereby completing the circuit.

The war had ended by the time the computer, called the ENIAC for Electronic Numerical Integrator and Computer, was completed in 1945. The machine contained 17,468 vacuum tubes, weighed 30 tons, and measured 18 feet high and 80 feet long, more than twice the size of Howard Aiken's Mark I.

The ENIAC manipulated numbers in decimal rather than binary form, using a simple counting method to add, subtract, multiply, or divide. In spite of its awkward computing method, the ENIAC was a stunning success. Because its switching was performed electronically, the ENIAC could speed through calculations a thousand times faster than the Mark I. Although the ENIAC had been supported by the Aberdeen Testing Ground for computing ballistics trajectories, its first job was to run a simulation of an early design for the hydrogen bomb. Already the versatility of the digital computer for solving problems of the physical sciences was becoming apparent.

As these advances proceeded in America, a gifted English mathematician, Alan Turing, was carrying out a far-reaching analysis of the universality of computing. Turing's work on the metamathematics of formal systems, such as logic, arithmetic, or computation, created the abstract notion of a computing system as a manipulator of the tokens in a formal system. Up to this time, computers had been primarily conceived of as "number-crunching machines," but Turing pointed out that they are equally capable of massive "symbol crunching."

Cryptology was the first practical implementation of Turing's abstract discovery. In

1943 the British built a massive computer, called Colossus, to break the secret German code generated by the "Enigma" machine. As a result of their success, which was a key factor in the outcome of World War II, the writing and breaking of codes became a powerful driving force behind the development of supercomputing. Because of the secrecy then, as now, surrounding this symbol-crunching side of supercomputers, very little of the impact of the intelligence community on the design and applications of supercomputers has yet been made public.

The Emergence of the von Neumann Architecture

John von Neumann was spurred to bring his fabled genius to the computing world after learning of the capabilities of the soon-to-be-completed ENIAC machine during an accidental encounter with Herman Goldstine at a railroad station. Von Neumann shortly began to collaborate with Eckert, Mauchly, and Goldstine in the hope of creating a supercomputer that could

Operators instructed the ENIAC by plugging in wires and setting switches; in effect, they had to reconfigure the computer's wiring every time the ENIAC had a new problem to solve. Reprogramming thus went at the slow speed of human hands.

simulate the complex physics involved in the design of nuclear weapons. The outcome of this association was one of the most influential papers in the history of computer science and engineering. The unpublished "Preliminary Discussion of the Logical Design of an Electronic Computing Instrument," authored by von Neumann, Goldstine, and Arthur Burks in 1946, contained a clear analysis of what became known as the von Neumann architecture, a vision of the computer that influenced its design for decades.

Von Neumann's archetypal computer would have a memory in which *both* data and operating instructions were stored. Because the computer's program resided in its memory, users could change instructions by rewriting the program instead of rewiring the machine. The computer would also have a central processing unit (CPU) connected to the memory by a bundle of wires often called a "bus." The CPU would consist of two parts: an arithmetic/logic unit that performs mathematical operations, and a central control unit that has a clock and orchestrates the computer's functions. Finally, the computer must have a device to accept data and display results, perhaps punch cards or magnetic tape or even a keyboard and a television monitor.

A von Neumann computer would perform a typical calculation in four sequential steps. The first step extracts a number from memory, the second extracts a different number from memory, the third tells the CPU to perform an arithmetic operation, such as addition or multiplication, on the two numbers, and the fourth returns the result to memory. Because the control unit requires these operations to be carried out in strict step-by-step order, the von Neumann computer is said to operate by scalar processing, from the Latin word *scala,* meaning "stairs."

At the time the historic paper was written, no true von Neumann computer with a stored memory existed. During the next few years, however, a whole sequence of machines that embodied these principles were created, with names like EDVAC and MANIAC. In particular, the Institute for Advanced Study at Princeton committed itself to building the "IAS" machine, one of the first binary computers, for von Neumann himself. In 1949, the University of Illinois undertook to build two copies of the von Neumann machine: the ILLIAC I for use at the Urbana campus and the ORDVAC for use at the Aberdeen Proving Ground. Today, the University of Illinois is in its fifth decade of heavy involvement with the world of supercomputing.

The clock in a computer's central processing unit allows us to compare the speed of computers. A single clock "period" is the shortest interval during which the computer can carry out an operation, like adding two numbers or storing a number. The clock period of a modern computer is almost inconceivably brief. For instance, the CDC 7600, which became the leading supercomputer of the early 1970s, had a clock period of 27.5 nanoseconds (a nanosecond is a billionth of a second), and a widely used modern supercomputer, the Cray Y-MP, has a clock period of 6 nanoseconds.

As the clock cycle shortens thanks to advances in electronic circuitry, computers can carry out more calculations per second. The rate of calculation provides a second way to measure the speed of computers, named for the format in which numbers are carried. Modern scientific computers represent numbers in powers-of-ten notation. For example, 1,363,857 is represented as $0.1363857000000 \times 10^7$. This arrangement is called floating point format because the decimal point moves or "floats" to ensure that all numbers have the same form. Modern supercomputers carry about 13 digits to the right of the decimal point in every floating point number. The speed of the computer can then be measured by how many floating point operations it can carry out every second, or how many "flops." For older computers that represented numbers in other formats, we can convert their speed into "equivalent flops."

Since the 1939 Mark I could perform a multiplication in 3 seconds, its speed is said to be 0.3 flops, whereas the 1946 ENIAC, which could carry out 400 multiplications per second, ran 1200 times faster at 400 flops. The CDC 7600 could run at 10 million flops (or 10 megaflops) by the late 1960s, and the national goal for the 1995 supercomputer is one trillion flops (or one teraflops). These figures give the so-called peak speed for the computer; in actual practice, scientists' applications run at only a fraction of this speed. However, it is convenient to use the peak speed when comparing a number of different types of computers.

The graph on this page shows the peak speeds of successive supercomputers during the last fifty years. Such a graph illustrates the steady increase in speed as mature technologies are replaced by newer ones, and the new technologies are then gradually exploited to their full potential. Three of these fundamental shifts in technology caused a rapid increase in a short period: relay to vacuum tube, vacuum tube to transistor, and today's shift to massively parallel machines made with hundreds to thousands of microprocessors. A fourth shift, from transistor to integrated circuit, caused a more gradual increase in power.

There are two fundamental reasons that basic innovation in electronic switch technology leads to a large increase in computing power. One is that the electronic components become

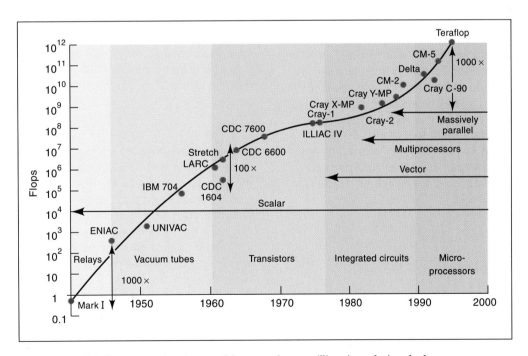

The speed of digital computers has increased by more than a trillion times during the last five decades. The exponential increase in speed (measured in flops) has continued at a steady pace even though the underlying hardware technology and the supercomputer architecture has undergone major shifts.

smaller and therefore many more can be packed into the same or smaller volume. For instance, the Mark I had 3000 relays, the ENIAC had 17,000 vacuum tubes, the Stretch had 170,000 transistors, the Cray-1's integrated circuits contained one million transistors, and the Thinking Machine Corporation's CM-5 delivered to Los Alamos in 1992 had 1024 microprocessors containing more than 500 million transistors altogether. By 1995, when scalable supercomputers will achieve a computational speed of more than one trillion multiplications per second, the transistor count will be well over 10 billion.

Not only does the total number of electronic switches continue to increase exponentially with time, but the fundamental switching speed of each switch decreases exponentially with time as well. The electromagnetic relays of the Mark I required 0.01 second to open and close, whereas the transistor in a 1995 microprocessor-based supercomputer will switch on or off in a few billionths of a second. The overall increase in speed from the Mark I to the 1995 supercomputer is therefore roughly the product of the increase in number of switches multiplied by the increase in speed of switching. The 1995 teraflop machine will have roughly three million times the number of switches of the 1939 Mark I, and each switch will act a million times faster. Thus one would expect a cumulative speed increase of three trillion, which is exactly the ratio of the speeds of the two machines.

Enter the Transistor

During the 1950s, an important innovation transformed the computing world: the vacuum tube was replaced by the transistor, a pea-sized electrical switching device invented by John Bardeen, William Shockley, and Walter Brattain at the Bell Telephone Laboratories in 1948. The

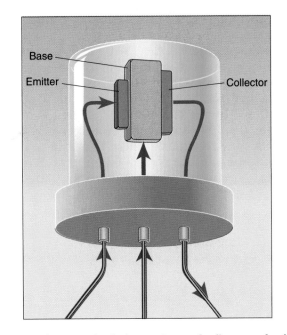

Early transistors consisted of an emitter and collector made of germanium or silicon "doped" with impurities to provide an excess of electrons. The base, which has a deficiency of electrons, is said to provide "holes." When a positive charge is applied to the base, electrons flow from the emitter to the collector.

transistor had several advantages over the vacuum tube: it was smaller, it was more reliable, it emitted much less heat. But most of all, a transistor could switch in much less time than a vacuum tube and therefore had the potential to increase the overall speed of the computer.

Once again the government intervened to accelerate the development of the next stage of computing technology. The two national nuclear weapons laboratories, Livermore and Los Alamos, challenged the private computer manufacturers to create an all-transistorized computer that would perform one hundred times faster than their best vacuum-tube supercomputer. The two largest computer manufacturers, IBM and Remington Rand, rose to the challenge.

Remington Rand had grown from the company that in 1951 created UNIVAC, the first commercially successful vacuum-tube computer.

Rand won the Lawrence Livermore contract and in four years had built LARC, the Livermore Advanced Research Computer. Finished in 1960, LARC with its 60,000 transistors was the first large computer to incorporate the new technology. Meanwhile IBM went to work for Los Alamos building the 7030, which was soon christened Stretch. It contained 169,100 transistors and was delivered to Los Alamos in 1961. The original contract had specified that Stretch should be one hundred times faster than IBM's 704, but the machine fell short of this goal, achieving an actual speed only about thirty times faster.

Although both machines were eventually delivered to the nuclear weapons laboratories, the two companies had each lost tens of millions of dollars on their projects and failed to produce commercially viable versions. At the same time, demand was escalating for machines that could merge and sort payrolls and corporate databases. Both companies decided to pursue the more lucrative and much larger data processing market, leaving a vacuum in the scientific and engineering numeric processing market. This vacuum was soon filled by a new company, the Control Data Corporation, and its young computer designer, Seymour Cray.

For the next thirty years, two largely distinct markets developed: one the large commercial business market, where the most powerful computers were called mainframes, and the other a small niche market for scientific computing, where the fastest computer was called a supercomputer. This hardware split, into the world of mainframes and the world of supercomputers, was paralleled by a split in the development of programming languages. COBOL became the most widely used language for business, while FORTRAN became the language of choice for scientists and engineers.

The Languages of Computing

To solve a problem or perform a simulation, a computer is instructed to follow a logical plan called an algorithm. The choice of an algorithm obviously depends on the nature of the problem under investigation, but it also depends on the computer's architecture. The software programmer who understands a supercomputer's architecture can pursue a strategy that exploits the machine's inherent computational strengths. Improvements in algorithm design are as important as advances in hardware in enhancing a supercomputer's overall performance. In Chapter 3 we shall return to a more detailed discussion of the types of algorithms used in supercomputing.

Once an algorithm is chosen, the programmer writes the specific steps to be followed in a language that the computer can understand. While the fundamental logic of program instructions has remained the same since Lady Lovelace's early insights, the practical implementation has evolved considerably. To program the ENIAC to perform a given computation, the operator had to set individually the several thousand dials on the front panel (each dial with ten possible settings) as well as insert the correct combination of patch cords hooking the various subcomponents of the computer together. With the adoption of binary numbers and the ability to store a program in the machine, the natural mode of programming became that of "machine language." In machine language, every instruction is a binary string of ones and zeroes that represents a set of commands (such as "fetch a number from a given location in memory and move it to the adding unit"). Machine language is an example of a low-level language.

As programmable computers were introduced in the early 1950s, demand began to develop for a language more appropriate for the human user. Grace Hopper, first with UNIVAC and later with the U.S. Navy, developed a translating program called a "compiler" that would take in higher-level, more English-like statements and convert them to machine code. By 1953, IBM had undertaken to create a general programming language for the IBM 704 that would be compiled. Under a team led by mathematician John Backus, IBM developed FORTRAN and released it by 1957. FORTRAN (for FORmula TRANslation) was immediately very popular because it widened the base of users who could become involved in programming.

Unlike machine language, FORTRAN used English words (such as *do* and *if*) and allowed for systematic management of DO loops and variables. FORTRAN's other great advantage was that, in principle, it could encode instructions for a wide variety of machine types. Thus, the user could write a FORTRAN program without having any specific knowledge about the hardware architecture on which the program would run. The compiler program, whose construction was invisible to the user and whose creation was the responsibility of the individual computer companies, took over the task of adapting general FORTRAN commands to the architecture of the machine.

This simple division of labor spurred a rapid growth in the use of computers. Within a few years, FORTRAN compilers were available for a variety of machines. The number of research projects involving computers expanded swiftly as scientists and engineers quickly learned the language. Shortly thereafter, the computer manufacturers realized that they needed to create a language focused on the needs of business data processing. By 1960, COBOL (for COmmon Business Oriented Language) had emerged for this purpose. Although many other computing languages have appeared in the intervening thirty years, COBOL and FORTRAN have remained the two dominant languages. Indeed, almost all of the examples of the use of supercomputers in this book were carried out by programs written in FORTRAN.

Seymour Cray's Supercomputers

Encouraged by the success in the early 1960s of its first, business-oriented computer, the new manufacturer Control Data Corporation set its sights on the scientific and engineering community that was embracing FORTRAN. Although the speed of its first product, the CDC 1604, was soon eclipsed by LARC and Stretch, the 1604 was fast, reliable, and reasonably priced. For slightly less than a million dollars, commercial customers could purchase a fully transistorized computer. The designer of the 1604 was Seymour Cray, a young genius who would personally lead the development of supercomputers over the next two decades. The introduction of his next computer, the CDC 6600, gave users a machine capable of performing up to 9 megaflops. Reflecting the split in the computer marketplace, the CDC 6600 is often regarded as the first commercially successful supercomputer, as contrasted with the data processing mainframe. It was followed in 1968 by the CDC 7600, also designed by Cray, which could race through computations at up to 40 megaflops. Both machines were enormously successful, and Control Data Corporation dominated the supercomputer business well into the 1970s.

The next major electronic advance was the replacement of single transistors with integrated circuits, invented in 1959 by Jack St. Clair Kilby at Texas Instruments. The elements of an integrated circuit, including multiple transistors and the interconnecting wiring, are inseparably associated in a self-contained module. The advantages of integrated circuitry were promptly exploited by two engineers, Jean Hoerni and Robert Noyce, at newly formed Fairchild Semiconductor. The resulting device came to be called a "chip," because its flat transistors and wiring are laid down on a single small slice of silicon. Within five years, primitive integrated circuits began to be used in commercial computers.

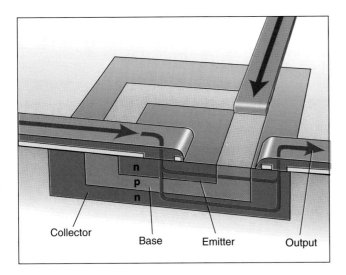

Collector   Base   Emitter   Output

A planar transistor, used in integrated circuits, is quite flat.
Jean Hoerni developed the so-called planar process, which
makes a transistor by depositing thin layers of semiconducting
material on a wafer of silicon. Robert Noyce realized that the
components of an integrated circuit could be connected by first
coating them with an insulating layer of silicon dioxide, then
etching small holes in the coating where electric contacts were
desired, and finally evaporating a thin layer of metal on the
wafer containing the circuit. This process allows thousands of
transistors, together with resistors and capacitors, to be
manufactured on a single silicon chip.

The advent of the silicon chip revolutionized logic circuitry and computer architecture.
Logic circuit boards occupying a square foot in
area shrank to a mere square centimeter. Electric impulses darting from switch to switch at
roughly half the speed of light traveled distances
of only hundredths of an inch. Power consumption and hence heat dissipation were reduced
dramatically. All of these technological improvements were potentially useful to the design of
supercomputers. However, unlike the transition
from relays to vacuum tubes or from vacuum
tubes to transistors, the move to integrated circuits did not cause a discontinuous jump in the
speed of supercomputers. Rather, it allowed for

the development of much more compact machines as the fundamental component of architectural design became not single transistors, but
functional units composed of entire circuits. The
introduction of the integrated circuit marked the
beginning of a period in which advances in
speed were enabled more by the design of the
supercomputer than by increases in the raw
speed of the components themselves.

Seymour Cray founded his own company,
Cray Research, in 1971. Five years later this
fledgling company unveiled its first supercomputer, the Cray-1. With this machine Seymour
Cray began to use integrated circuits, although,
in his characteristic demand for reliability, the
scale of integration was far from the state of the
art at the time. More important than the level of
integration was his choice of an architectural
design that used integrated circuits to boost the
performance of his machine. While the Cray-1's
clock cycle of 12.5 nanoseconds was a little less
than half that of the CDC 7600, its peak speed
of 170 megaflops was over four times faster than
the peak speed of the CDC 7600. The additional speed of the Cray-1 beyond that expected
from just comparing clock cycles can be attrib-
uted to Seymour Cray's implementation of a
recent architectural innovation called vectoriza-
tion.

Architectural Innovations

A successful computer architect will design a
computer's circuitry in a way that is advanta-
geous to the computer's eventual use. Seymour
Cray improved the performance of the Cray-1
by exploiting a new architecture that was better
adapted to the structure of the programs it normally ran.

Quite commonly, a particular mathematical
procedure such as the multiplication of two
numbers must be repeated many times in a program. For instance, suppose you have two sets

of a hundred numbers—call them As and Bs—that must be multiplied in pairs. A program written in FORTRAN would accomplish the repeated multiplication with a DO loop:

```
    DO 10 I = 1, 100
    C(I) = A(I)*B(I)
10 CONTINUE
```

The computer first multiplies A(1) and B(1), the first two numbers in each set, and stores the results as C(1), then it multiplies A(2) and B(2) and stores the result as C(2), and so forth until it multiplies A(100) and B(100) to produce C(100). After storing this final multiplication, the computer moves to statement number 10, which tells the computer to continue on to the next instruction in the program.

Arithmetic is performed in circuitry called functional units, and the multiply functional unit in a typical supercomputer like the Cray X-MP has seven segments. Steering modules at the input and output of the functional unit direct numbers into and out of the unit. Thus a single multiplication takes nine clock periods: one clock period for the numbers to enter the input steering model, seven clock periods to move through the segments of the functional unit, and a final clock period for the result to leave through the output steering module.

A scalar processor designed according to the classical von Neumann architecture requires the computer to complete one multiplication before the next multiplication is begun. To multiply a hundred pairs of numbers therefore requires 900 clock periods. Only one multiplication occurs at a time, and thus most of the circuitry is idle as a single pair of numbers proceeds along its journey through the functional unit. Scalar processing is characterized by this "von Neumann bottleneck"—the inability to work on producing more than one result at a time, analogous to a single laborer manufacturing a product from beginning to end.

A significant improvement over the traditional von Neumann scheme is pipelined vector processing. This scheme can be compared to an assembly line. A vector processor feeds a new pair of numbers into the steering module and from there into the functional unit hard on the heels of the previous pair of numbers. Pipelining was originally introduced in scalar form by IBM. It ensures that all segments of the functional unit are working on numbers at some stage of execution. Once the vector pipeline is filled, the CPU produces one result for every clock period.

To multiply a hundred pairs of numbers, for instance, a vector processor requires only 109 clock periods: 9 clock periods to fill the pipeline (called the startup time) plus 100 clock periods to obtain the one hundred answers. Vector pipelines had previously been introduced in 1974 by both CDC, in the STAR 100, and by Texas Instruments in their ASC supercomputer.

Of course, vector operation can work only if the problem one is solving has a naturally well organized data set. For problems in some areas, such as fluid dynamics with its neat grids, data sets are easily arranged for vector processing; but in areas such as quantum chemistry, where one may be computing only a few atoms, the problems are not easily "vectorized." As a result, a vector supercomputer will have a much larger *range* of performance times on a broad set of applications than the strictly scalar von Neumann machines that had come before.

For the last twenty years the supercomputing community has had to learn to live with this phenomenon of a single machine having a large range of performance. For instance, to achieve the peak speed of 330 megaflops on a Cray Y-MP vector processor, the user's code must be 100 percent vectorized. However, the average speed achieved by NCSA users is only 70 megaflops. Although some codes routinely sustain speeds as high as 220 megaflops, others run at only 10 megaflops.

Most users are resistant to new architectural designs, for good reason. It is daunting to have to rewrite a computer program to take advantage of the new innovation, especially when that program may be the culmination of years of development. Furthermore, the new features are often poorly supported by the computer's software environment for years after the hardware is running perfectly. Seymour Cray evidently understood this truism better than some of his competitors. Although he had simplified the vector hardware in his first machine, the Cray-1, more crucially he had also made a special effort to increase the scalar speed of the machine. By doing so, he ensured the rapid market acceptance of the Cray-1 and the rejection of the other major vector machine, the STAR 100. By more than doubling the scalar speed of the Cray-1 over that of the reigning CDC 7600, Seymour Cray made certain that *all* jobs would run faster, whether or not they made use of the new vector feature. Thus the scientific community could immediately enjoy swifter computing while taking several years to learn how to reprogram their codes to take advantage of the faster vector units.

A Supercomputer Famine in the Universities

Ironically, just as faster computers for scientific research were being introduced, they were becoming less available to the scientific community. For a variety of complex reasons, the entire profile of U.S. government funding for science abruptly changed in 1970. Throughout the 1960s, colleges and universities in the United States had enjoyed generous growth in science funding, partly in response to the 1957 launch of *Sputnik*. But by 1970 the expense of the Vietnam War and its inflationary effects on the economy had induced the government to substantially cut increases in science funding to universities, whose tolerance of vociferous resistance to the war had not endeared them to government leaders. As a result of these and other factors, by 1976 American universities were annually graduating only half the number of Ph.D.s in engineering that were graduated in 1969.

As the growth rate of government funding declined, large-scale university computing facilities were a prime casualty. As an example of the effects of the cuts, no American university was ever able to purchase a CDC 7600 during its long reign as the undisputed supercomputer. It was not until 1985, with the formation of the National Science Foundation's supercomputer centers, that the government reinstated its traditional partnership with universities and computer manufacturers.

After the federal government reduced its support, most of the university community retreated from the front lines of supercomputing. Most work was performed on minicomputers, particularly those manufactured by the Digital Equipment Corporation (DEC). Indeed, if it had not been for the DEC VAX computer, the university community's use of computers to support science and engineering research might have ceased altogether.

During the travail of the 1970s, the nuclear weapons laboratories at Los Alamos and Livermore remained relatively immune to funding cuts. As the chart on the next page illustrates, these laboratories were able to assemble whole fleets of vector supercomputers. Indeed, their software engineering teams were instrumental in developing much of the software that made these computers useful. These laboratories, which had been using supercomputers intensively for forty years, became the national repository of supercomputing expertise. Scientists there had the best understanding of how to use high-performance computers to solve complex problems in science and engineering as well as how to create and manage a comprehensive facility for supporting such research.

In addition to secure facilities for nuclear weapons design and for intelligence activities, there were several open national centers for research only in special topics such as atmospheric science or geophysical fluid dynamics. University researchers in other disciplines who wanted to carry out supercomputing science did not, in general, have access to these federal facilities. A

Computer	Power (Cray-1 = 1)
RR UNIVAC-1	0.000005
IBM 701	0.0003
IBM 704	0.0005
IBM 704	0.0005
IBM 704	0.0005
IBM 709	0.0005
IBM 709	0.0005
IBM 7090	0.003
RR LARC	0.008
IBM 7090	0.003
IBM 7030	0.016
IBM 7090	0.003
CDC 1604	0.002
IBM 7094	0.004
IBM 7094	0.004
CDC 3600	0.008
CDC 3600	0.008
CDC 6600	0.050
CDC 6600	0.050
CDC 6600	0.050
CDC 7600	0.225
CDC 7600	0.225
CDC 7600	0.225
CDC Star	0.450
CDC Star	0.450
CDC 7600	0.225
Cray 1	1.0
Cray 1	1.0
Cray 1	1.0
Cray 1	1.0
Cray 1	1.0
Cray 1	1.0
Cray XMP/2	2.5
Cray XMP/2	2.5
Cray XMP/4	5.0
Cray 2	5.0
Cray XMP/4	5.0
Cray 2	5.0
Cray YMP/8	15.0
Cray 2	10.0
Cray XMP/1	1.2
Cray C-90	50.0

The history of large computers at Lawrence Livermore National Laboratory illustrates the shifting dominance of a few vendors: first IBM, then Control Data Corporation, and most recently Cray Research. The Cray Research C-90 arriving at LLNL in late 1992 will be 10 million times faster than the laboratory's first supercomputer, the UNIVAC.

few of us were able to obtain security clearances and spend the summers at the national laboratories, but the rest of the year we had access only to less advanced university facilities.

By the early 1980s, some European countries had placed supercomputers in open research laboratories. A number of American computational researchers began to spend their summers abroad in order to pursue their research using American-built supercomputers. The world of today—in which thousands of researchers in hundreds of universities routinely access, over a high-speed national network, the latest in supercomputing equipment from the comfort of their office—was at best a distant dream.

Even as the supercomputer famine raged, the seeds of self-correction were being sown. The support of the government labs and certain large-capitalization industries was adequate to enable Cray Research to drive ahead, creating ever-faster supercomputers. The computer science and engineering community continued to explore a wide variety of new parallel architectures that held the potential for creating new kinds of supercomputers. Finally, the university scientific community began to realize that it must regain access to the fastest computers. The federal government itself began a series of studies that reached the same conclusion. These renewed efforts led to the formation of the National Science Foundation supercomputer centers in 1985 and the creation of the national network in the late 1980s.

The Beginnings of Parallelism

In the late 1960s an important innovation in computer design came from a University of Illinois team led by Dan Slotnick. The Illinois computer scientists and engineers designed a machine, the ILLIAC IV, that once built by the Burroughs Corporation became the first massively parallel computer. In contrast to the STAR 100, ASC, and Cray-1, with their single vector units, the ILLIAC IV had 64 identical scalar computers that operated in parallel, each with its own processing unit and local memory.

In a parallel computer, a particular computation is divided among the processors, which complete the job much faster than a single CPU laboring in isolation. Suppose, for instance, that 100 pairs of numbers are to be multiplied on a parallel machine with two processors. One processor might calculate the multiplication of the first 50 pairs of numbers while the other processor calculates the multiplication of the other 50 pairs. As the calculation proceeds, the processors march forward in lock step, so that they all fetch from memory, add a pair of numbers, and return to memory simultaneously. Like the vector unit, the parallel computer achieves consid-

The five-foot cube houses the 65,536 processors of the CM-2 Connection Machine, designed by Danny Hillis of Thinking Machines Corporation. The curved structure on the right is the mass storage system, whose dozens of small disk drives provide rapid access to 10 gigabytes of data.

erably more speed than a strict von Neumann architecture.

Thus, on a parallel machine, a program is run concurrently on many processors. This requirement taxes the programmer, who must figure out how to distribute the workload evenly among the processors. Although its cost of more than 100 million 1990 dollars makes the ILLIAC IV one of the most expensive supercomputers ever built, it did not lead immediately to a market product. Its basic architecture was reborn in the Connection Machine when the Thinking Machines Corporation was founded ten years later in 1983.

A Supercomputer Is a CRAY

Even before the Connection Machine arrived on the scene, Cray Research had developed a different form of parallelism. The Cray-1 was a wonderful machine that fired the imagination of a whole generation of computational scientists and engineers. Building on this early success, Cray Research introduced in 1982 the first vector multiprocessor, a vector machine having up to four processors, each with a clock period of 9.5 nanoseconds. This machine was the Cray X-MP supercomputer, whose parallelism was added by Steve Chen, who obtained his Ph.D. from the University of Illinois. It could hurtle through calculations at speeds as high as 860 megaflops if the problem was run in parallel.

The Cray Research X-MP was followed in 1985 by another four-processor machine, the Cray-2. This compact machine broke the gigaflop barrier—it could execute in excess of one billion floating point operations per second. Moreover, its memory was 16 times larger than that of the X-MP. Although each of the 240,000 chips in the Cray-2 gives off only a tiny amount of heat, they are packed together so tightly that

The Cray-2 is filled with a colorless, odorless, inert fluorocarbon liquid that continually flows across the circuitry and then through the elegant fountain standing behind it. About a third of the computer's 5500-pound weight consists of coolant. The compact machine is composed of 14 vertical columns arranged in a 300° arc, 135 centimeters in diameter and 115 centimeters high.

their combined heat output poses a significant threat to the machine's circuits. Cray Research engineers solved this problem by immersing the entire circuitry in a fluorocarbon liquid that continually circulates through the machine. The liquid flowing through the sleek, transparent modules gives the Cray-2 a surrealistic appearance akin to a beautiful piece of sculpture.

In 1987, the Cray Research Y-MP emerged, offering up to 8 processors, each with a peak speed of 330 megaflops, for a parallel peak speed of 2.6 gigaflops. As this book is being written in 1992, the Y-MP is the standard for what most people mean by supercomputer. Most of the science examples in this book were performed on a Y-MP or a Cray-2 supercomputer. The successor to the Cray Research Y-MP, the Cray C-90, which has 16 processors and a factor of two increase in the clock cycle, is just coming to market at the end of 1992. During this amazing sixteen-year period of dominance by Cray Research, speed has increased a hundred-fold from the first Cray 1 in 1976 to the parallel C-90 in 1992. Even more spectacular has been the growth in memory size, by a factor of 500.

In many ways the increase in memory is more important than the increase in speed. If a scientist cannot fit the computer program and all the arrays of data into the memory, then he or she must lessen the complexity of the model or its spatial resolution and therefore learn less about the science. The memory of most personal computers in 1992 is a few megabytes

(1 byte equals 8 bits), almost as large as the 8-megabyte central memory of the 1976 Cray 1. Once again we see how transient the label "supercomputer" is. The 16-processor C-90 from Cray Research will have over 4000 megabytes of memory, while the 1995 teraflop supercomputer may have as many as one million megabytes (or one terabyte) of memory.

Why Speed Matters

Improvements in speed and memory have a significance that can be appreciated by comparing images from identical computer programs run on machines of differing power. In 1982, one of us, Larry Smarr, and graduate student John Hawley, both of the University of Illinois, were studying how gas flows onto a black hole. Together with colleagues Robert Wilhelmson, Richard Crutcher, and Robert Haber, we had set up an interdisciplinary VAX and Image Processing center at Illinois. An overnight run on the DEC VAX minicomputer, as powerful a machine as most universities had at that time, would yield a solution to the general relativistic gas flow equations and create a visualization of that solution.

The first image on the next page shows the typical output from such a run. The simulation is symmetrical around the vertical axis on the left of the diagram, placing the center of the black hole in the lower left corner. Gas originally flowing inward from the right spirals around the central black hole until it forms a torus, whose maximum density lies at the center of the series of contour lines shown in the diagram. One can easily see the resolution of the discrete grid, since each of the tails of the yellow arrows denoting the velocity of the gas is tied to one corner of a grid zone. The VAX simulation had convinced us that these orbiting gas tori could form, but even on an eight-hour, over-

The newest version in the long line of Cray Research supercomputers is the C-90, which can have up to 16 processors and 4 gigabytes of memory. Each processor has a peak speed of one gigaflop; when all 16 processors are computing in parallel, the machine has a peak speed of 16 gigaflops.

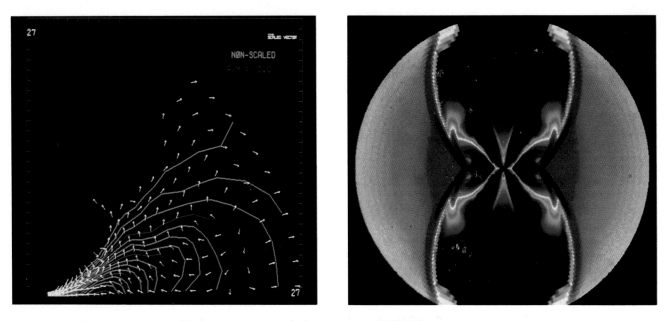

A simulation of gas flowing onto a black hole, from an eight-hour run on a VAX (left) and a three-hour run on a Cray-1 (right). The run on the Cray-1 produced an enormously higher resolution on a much larger volume of space.

night run we couldn't compute enough grid zones to let us see what was happening in detail.

By traveling to Munich in the summer to work with our colleagues Karl-Heinz Winkler and Michael Norman, we were able to run our computer program on the first Cray-1 in continental Europe, at the Max Planck Institute for Physics and Astrophysics. The program ran four hundred times faster! Our previous eight-hour run now took only one minute. Therefore we could run the simulation overnight with much finer zoning on a larger physical space. As a result, we avoided artificial complications produced because our computational boundaries were too close to the black hole. The resulting visualization is also shown on this page.

The difference between the two images is stunning. By placing the computational outer boundary much farther from the hole, we could allow the gas to fall freely toward the black hole until the centrifugal force began to decelerate the spiraling gas. The sudden deceleration created the beautiful large-scale standing shock waves that form a figure eight. The gas flow abruptly changes direction after passing through the shock and "slides" down the inside of the shock, nicely funneling into the growing torus. Both runs used exactly the same computer program, but the speed of the supercomputer allowed us to bring the physics sharply "into focus," by using many more grid points in the simulation. It was as if after being severely nearsighted all our lives we had put on a pair of glasses!

The ability of faster and faster supercomputers to reveal finer and finer details of gas flow is not just of academic interest. One of the attributes that aeronautical engineers explore in their analysis of an aircraft design is the aircraft's behavior during a turn, when the aircraft is tilted to its direction of motion through an angle called the angle of attack. At moderate to high angles of attack, air flow separates from the leading edge of an airplane's wing and forms a pair of spiral vortices above the wing's upper surface. Low pressure induced on the upper surface by these vortices gives the delta-shaped wings of a fighter aircraft added lift. At large attack angles, however, the vortices can degenerate into a chaotic swirling pattern, causing a sudden loss of lift. This critical phenomenon, called vortex breakdown, was well known from wind tunnel experiments, but until recently aeronautical engineers did not have computers able to perform numerical simulations at a resolution high enough to reveal the breakdown.

Three attempts to simulate vortex breakdown by solving the equations of fluid flow on a grid surrounding the swept wing are shown on the next page. A numerical simulation possible in 1978 on a CDC 7600 could solve the equations at 36,000 grid points and fails to show the breakdown. A simulation in 1984 using a Cray X-MP and 120,000 grid points reveals the flow in greater detail but still shows no breakdown. Finally, in 1986, a Cray-2 had the speed and memory to solve the equations at 800,000 grid points. At last the resolution was sufficient to reveal the vortex breakdown clearly. This success has given aeronautical engineers important clues that may lead to the eventual control of vortex behavior.

The Rise of the Microprocessor

Despite the great market success of multiple-processor vector supercomputers, few users were taking advantage of their peak processing power. Since there were typically more users waiting in the queue to use the supercomputer than there were processors, it did not seem reasonable to

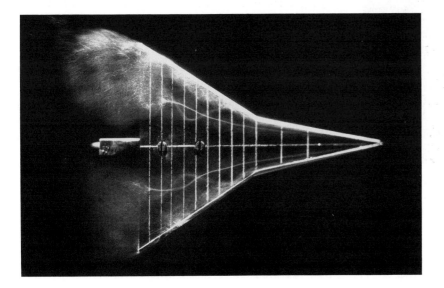

Modern fighter aircraft obtain extra lift from swirling air flow patterns generated by highly swept wings and strakes (narrow extensions in front of the wings). This picture from a wind tunnel experiment shows vortex breakdown near the rear of an aircraft's delta-shaped wing.

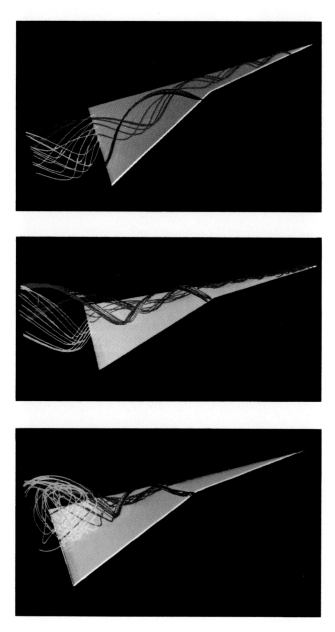

The three images, which are the results of numerical simulations of air flow over the delta-shaped wing of a fighter aircraft, show the advantage of increased resolution made possible by advances in supercomputer technology. The lines in each picture show paths traced by hypothetical particles in the air flowing past a wing on one side of the aircraft. Vortex breakdown appears only in the 1986 simulation, as jumbled yellow particle tracers.

give all the processors to one user and make everyone else wait until that one computation had finished. Therefore, most scientists still ran their calculations on only a single one of the available processors. In comparison to the potential four- or eight-fold speedup offered by a multiprocessor, the single-processor users were experiencing a very slow growth in computing capability. For instance, it took about a decade to halve the clock cycle from 12 nanoseconds for the Cray-1 to 6 nanoseconds for the Cray Y-MP.

While the single processors in vector machines have been eking out small increments in speed, microprocessors have been gaining speed exponentially. Invented by the Intel Corporation in 1971, the microprocessor is an integrated circuit that has all the components of a full computer. This "computer on a chip" formed the basis of the personal computer revolution that started in the late 1970s. Although only a few hundred sites own supercomputers, there are tens of millions of installed personal computers. Competition for this vast market drives intense efforts to improve microprocessor technology.

In contrast to the single processors in vector machines, microprocessors double in speed every eighteen months. During the same period that the Cray Research single processor doubled in speed only once, microprocessor speeds increased by a factor of about 64. Therefore, even though the Cray machines started out dozens of times faster than individual microprocessors in the mid-1980s, by 1993 a single Cray C-90 processor will have a clock cycle only slightly faster than the speediest microprocessors.

The first generation of microprocessors contained all the machine instructions needed by the computer to run the software, just as the "supercomputers" of the 1950s and early 1960s had. However, as CDC had realized in the mid-1960s, by devoting a larger fraction of the silicon chip to instructions for numerically intensive tasks, a manufacturer could create machines that handled these tasks much faster.

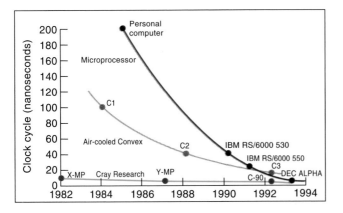

There has been enormous progress in the clock speed in microprocessors and mid-range machines over the last twenty years, but the clock cycle has changed much more slowly in single-processor vector supercomputers.

Indeed, it was at IBM that a team led by John Cocke in 1971 realized that one only had to include a small portion of the full instruction set in the microprocessor; the much less frequently used instructions could be placed in software. Again, the result was a computer that was much faster for scientific computing. This idea of a Reduced Instruction Set Computer (RISC) gradually took form in workstations introduced by IBM, Hewlett Packard, and Sun Microsystems. At first, people thought of RISC technology as having little to do with supercomputing, because these workstations, although faster than those built on Complex Instruction Set Computers (CISC), were still much slower than supercomputers. Furthermore, supercomputing had become synonymous with vector processing, so the scalar computing of microprocessors seemed to be a throwback to earlier times.

With the introduction in 1990 of the IBM RS/6000 family of RISC workstations, this perception began to change. The fastest IBM RISC microprocessors had a clock speed only one-quarter that of the reigning supercomputer, the Cray Research Y-MP. That meant that each IBM workstation was at least as fast as the CDC 7600. However, the CDC 7600 could execute only one instruction at a time, whereas the IBM RISC workstation is actually a "superscalar" machine that can execute several instructions per clock cycle. This extra feature allows the IBM workstation to rival a Cray Research Y-MP in speed on some applications that can take advantage of the superscalar architecture. On the other hand, if an application is highly vectorized, the Cray will still run much faster than the workstation.

These RISC workstations had up to 128 megabytes of memory, an amount enormously larger than the central memory on the CDC 7600 and indeed comparable to or even greater than the memory share of a single processor on a multiprocessor Cray Research Y-MP. Furthermore, some RISC vendors, such as Silicon Graphics, have begun to offer multiprocessor workstations that, like the Y-MP, have up to eight processors sharing the same memory.

But the RISC processors posed an even broader challenge to the dominance of supercomputers. Most physics departments or laboratories had many such workstations hooked together on local area networks. By 1992, it was becoming apparent that one could hook these clusters of workstations together with software and allow them to attack computational science or engineering problems in parallel. This loosely clustered system of microprocessors then potentially becomes a parallel computing engine capable of supercomputer performance.

Massively Parallel Supercomputers

A few years earlier, designers had carried the notion of hooking microprocessors together one step further. In their Cosmic Cube experiments of the early 1980s, the Caltech group of Charles Seitz and Geoffrey Fox had connected 64 microprocessors in a "hypercube" topology. A hypercube architecture of 2^n processors connects every processor to its n nearest neighbors by picturing each processor as the corner of an n-dimensional hypercube. Just as in three dimensions a cube has eight corners, each connected to its three nearest neighbors, so an n-dimensional hypercube has 2^n corners, each connected to n nearest neighbors. Fox and Seitz's machine captured a large fraction of the potential 64-fold increase in speed over a single microprocessor. Intel brought out a commercial version of this machine called the iPSC in 1985 and has continued to expand its line of products.

nCUBE has built a physical implementation of a 10-dimensional hypercube containing 2^{10} or 1024 processors. You might suppose that this supercomputer should be able to whiz through calculations 1024 times faster than a single processor, but a doctrine called Amdahl's law predicts that any given problem run on a parallel system will use the processors less efficiently as the number of processors increases. This skeptical belief, put forth in 1967 by Gene Amdahl, argues that the speedup one can obtain by adding more processors is limited by the eventual development of bottlenecks. By analogy, you can speed up office work by hiring more typists, but somebody has to dictate letters, draft reports, collate and staple the typed pages, and so forth. These activities require sequential steps that inevitably create bottlenecks that prevent you from doubling the work output by simply doubling the number of typists.

Scientists at Sandia National Laboratories circumvented Amdahl's law with a simple strategy. Instead of freezing the problem size as they increased the number of processors, they scaled up the size of the computational work load in proportion to the number of processors. This tactic is like giving each secretary a complete report to type rather than a few pages of the

The nCUBE 2 is typical of massively parallel computers in that it can be scaled up by adding more boards, one of which is shown here. Each board contains 64 identical computing elements, each with its own floating point unit and local memory. A single chassis can hold 16 of these boards, creating a 1024-processor system. Up to eight of these chassis can be hooked together to form an 8192-node supercomputer with a footprint smaller than 50 square feet.

same report that everyone else is working on. Of course, to take advantage of this strategy you must have more than one report to be typed. Bottlenecks still develop, but the computer is handling such a massive throughput of computation that the delays are comparatively small. In representative simulations involving wave propagation, mechanics, and fluid flow, Sandia scientists achieved a speed increase of between 1010 and 1020, remarkably close to the maximum of 1024.

The Sandia result leads to two generalizations that are important for thinking about the massively parallel world of the next decade. The first is a concept called "data parallelism" mainly elaborated on by Danny Hillis, the brilliant young computer architect of Thinking

Machines Corporation. As we saw in the thunderstorm example, the equations that we seek to solve remain constant as we change the number of grid zones by factors of hundreds to thousands. Thus there is much more parallelism in the "data" than there is in the equations. To increase the resolution on a computation, one can add more grid zones by adding more processors. One can imagine accumulating processors in a massively parallel system almost indefinitely as one's appetite for accuracy in describing nature grows. Thus there is a deep mapping between processors and the physical "parallelism" of nature itself. From this point of view, massively parallel machines are more "natural" for modeling large numbers of interacting physical elements, such as large biological molecules or colliding galaxies that consist of thousands of stars.

Proponents of the second concept, "scalability," envision building up massively parallel computers out of the same components used in desktop computers. Thus the user can scale up the number of processors as appropriate to the problem at hand without the software environment changing. One can imagine similar computing devices of many different sizes distributed throughout a "scalable" computing world of extremely portable software. One frontier of architectural design lies in discovering the ultimate limits of scalability both in hardware and in algorithms and software.

With the psychological barrier of Amdahl's law broken, many researchers believe that parallel processing is clearly the wave of the future in supercomputing. In 1990, Intel Corporation introduced the iPSC/860, the first in a series of parallel supercomputers based on the i860 RISC processor. The iPSC/860 comes with up to 128 processors capable of delivering a 7.6-gigaflop performance at a base price of only $3.5 million. In 1991 a consortium of universities led by Caltech stunned the supercomputing community by purchasing an Intel "Delta" machine containing 528 nodes, each one an Intel i860 RISC micro-

processor. The peak speed of this machine was an astounding 32 gigaflops—well over 10 times the peak speed of an eight-processor Cray Research Y-MP when it is used in parallel.

The comparison of the Delta machine to the Y-MP is not entirely fair since, as is always the case when new architectures are introduced, little software existed that could run the machine to its best advantage. For the vast majority of users, the barrier of having to learn a whole new way of computing still outweighed the potential speedup or cost savings. However, for the pioneers always willing to bear the pain of a rough new system in exchange for added performance, this announcement had an impact similar to CDC's announcement of the 6600 in 1964, which had broken IBM's previous dominance, or Seymour Cray's announcement of the Cray-1 in 1975, which had broken CDC's previous dominance.

One of the challenges of building parallel machines is synchronization: making sure that each processor has all the information it needs when the time comes for that processor to execute an instruction. One way of ensuring synchronization is to have a control processor broadcast identical instructions to all the other processors, and to have each of the processors then apply the instructions to data in its local memory. This scheme is called SIMD for "single instruction, multiple data." The Thinking Machine Corporation's CM-2 introduced in 1987 is a SIMD-supercomputer.

An alternative approach is to allow each processor to execute its own, separate program on the data in memory. In this mode, called MIMD for "multiple instruction, multiple data," the supercomputer must provide explicit means of synchronization. Processors exchange messages containing needed data, which also tell them when to wait or proceed, depending on the progress of all the other processors. Both vector multiprocessors, such as those from Cray Research or Convex Computer, and massively parallel machines, such as those from Intel and nCUBE, have MIMD architectures because each of the processors can carry on computations independently of each other.

Within months of the announcement of the Delta machine, an equally stunning announcement came from Thinking Machines Corporation: SIMD and MIMD architecture would be combined in the CM-5, shown on the first page of this chapter. Each of the CM-5's hundreds of nodes would be a RISC microprocessor identical to that in a Sun Microsystems workstation, but each node would be coupled to vector units that could, in principle, raise its computing power by a factor of 25. Each node of a CM-5 would have one-third the peak speed of a single Y-MP processor! Furthermore, in 1992 Los Alamos National Laboratory took delivery of a 1024-processor version of this machine, having a peak speed of 128 gigaflops and a memory of 32 gigabytes. Several other national centers simultaneously ordered smaller versions.

There is a significant difference between the current generation of vector machines and the massively parallel machines: the former possess a common memory shared by all processors, while the latter have a separate memory for each processor. These two classes of architectures are called shared and distributed memory, respectively. In general, it is easier to program for a shared memory, but at present the users of shared memory who want to work in parallel are limited to a small number of processors.

The massively parallel computer from Kendall Square Research introduced in 1992 represents the first of a new generation of machines that have both shared memory and large numbers of processors. In principle, it should be easier to move current application codes to these shared memory architectures. Both Cray Research and Convex Computer have announced machines of similar architecture for 1993.

There is both advantage and disadvantage for the users in this explosion of different architectures. On the one hand, users have a much greater chance of finding a hardware type that suits the character of their particular application. On the other hand, they could spend all their time rewriting their code to fit each new machine. An effort being led by Ken Kennedy, director of the NSF Science and Technology Center for Research in Parallel Computing at Rice University, to produce a portable parallel version of the FORTRAN language, supported by all vendors of parallel architectures, offers a possible way out of this dilemma. The users would simply write their code in this new portable FORTRAN, and it would run on any of the vendors' computers. It would be up to the vendor to develop a compiler that would map the user's code onto the particular parallel or vector hardware produced by that vendor.

Simultaneously with the entry of these new massively parallel companies, the market for vector multiprocessors is becoming more crowded. Designers Seymour Cray and Steve Chen each left Cray Research to form their own competing companies called Cray Computer Company and Supercomputer Systems, Inc. respectively. Both companies are expected to deliver vector multiprocessors in the mid-1990s. In 1991, Convex Computer introduced the first supercomputer using gallium arsenide instead of silicon. Its main advantage is that gallium arsenide generates less heat than silicon, thus allowing the computer to be air cooled. The Convex C3 also allows for a large shared memory—up to 4 gigabytes.

IBM meanwhile has sold hundreds of vector units for its 3090 line of mainframes, the largest of which has six processors. Its newest version of this vector mainframe, the 9000 series, is being delivered in 1992 with a peak speed of 2.7 gigaflops. Japanese supercomputer manufacturers in 1992 are delivering both highly optimized vector supercomputers—such as the NEC SX-4, which attains a peak speed of 22 gigaflops using only four processors, and Hitachi's S-3800-480, which has a peak speed of 32 gigaflops—as well as general-purpose mainframes, such as the Fujitsu 2600.

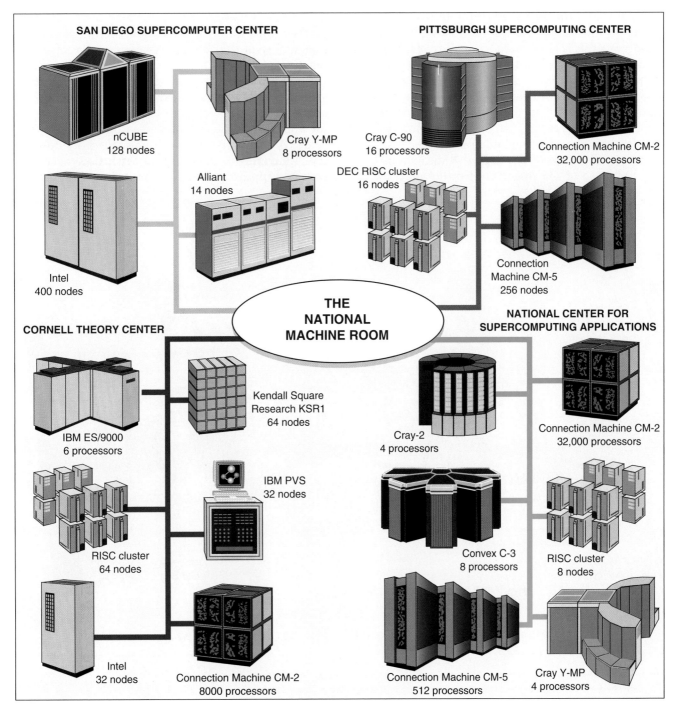

SAN DIEGO SUPERCOMPUTER CENTER

nCUBE
128 nodes

Cray Y-MP
8 processors

Alliant
14 nodes

Intel
400 nodes

PITTSBURGH SUPERCOMPUTING CENTER

Cray C-90
16 processors

Connection Machine CM-2
32,000 processors

DEC RISC cluster
16 nodes

Connection
Machine CM-5
256 nodes

THE NATIONAL MACHINE ROOM

CORNELL THEORY CENTER

IBM ES/9000
6 processors

Kendall Square
Research KSR1
64 nodes

IBM PVS
32 nodes

RISC cluster
64 nodes

Intel
32 nodes

Connection Machine CM-2
8000 processors

NATIONAL CENTER FOR SUPERCOMPUTING APPLICATIONS

Cray-2
4 processors

Connection Machine CM-2
32,000 processors

Convex C-3
8 processors

RISC cluster
8 nodes

Connection Machine CM-5
512 processors

Cray Y-MP
4 processors

*High-speed networks link together the high-performance computers at the four National
Science Foundation supercomputer centers to form a virtual National Machine Room.
Shown are the machines available to the national research community as of early 1993.*

Although the vector multiprocessor computers are no longer the fastest machines or those with the largest memories, they will still be considered supercomputers for some years to come. They will retain their status as long as most third-party application codes, such as many of those used in the examples in this book to perform engineering, chemistry, or biology simulations, are only available on the vector machines. Because of the high risk and small installed base of the massively parallel machines, it will take several years to develop a set of application software for these new architectures that will rival that already available on vector machines.

The number of potential customers is not growing nearly as fast as the number of supercomputer manufacturers. Thus, the stage is set for a classic market shakeout within the next few years. A company must sell enough of its current machines to generate the cash flow necessary to support the very expensive research and

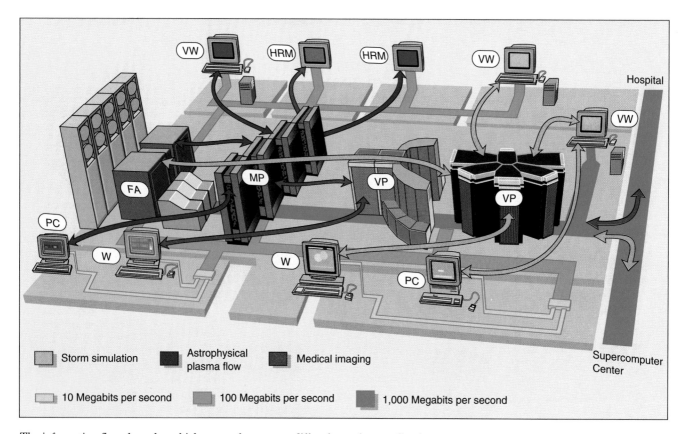

The information flow through multiple types of computers differs for various applications. One can think of a distributed set of computers as being one integrated metacomputer. The "garden of architectures" shown here is typical of a modern high-performance computing center.

development needed to create the next generation machine. With the shortening of the development cycle brought about by the conversion to RISC technology and the large growth in the number of vendors, it seems inevitable that several of these companies will be forced either into bankruptcy or into mergers within a few years. Economic factors will be at least as important as the elegance of the architectural designs in determining which companies are still thriving by the time the first teraflop machines are delivered.

A National Metacomputer

It would be almost impossible to take full advantage of today's proliferation of supercomputers were it not for personal computers and a national network that connects them to supercomputing centers. Scientists can develop their computer programs on personal computers at their desks, then run them on supercomputers at major computing facilities, and then analyze the data at their desks again. In the past, such facilities usually had only a single supercomputer, but today they retain older machines alongside those incorporating the latest innovations in architecture. By linking these machines through high-speed networks, the National Science Foundation centers have created a virtual National Machine Room, in which almost all commercially available supercomputers are placed within reach of the national community.

At the computing facilities, these same highspeed networks allow the weaving together of a variety of specialized machines to form a "metacomputer," or a virtual computer whose subcomponents are actually full-scale computers. The illustration on the facing page shows the data flows in a modern supercomputer center moving between vector machines, massively parallel machines, graphics computers, and mainframes. Note that each user of a metacomputer electronically connects the underlying computers in a form appropriate for the application at hand. When the program is finished, whatever files the user wishes to keep are transferred to the "mass storage" device, which is usually a mainframe equipped with both hard disks and tape. Typical 1992 supercomputer facilities hold over one million user files, containing many millions of megabytes of information.

With the advent of the National Machine Room, the metacomputer is rapidly becoming a reality on a national scale. However, unlike a personal computer whose multiple computing chips, buses, and storage devices were assembled by a vendor, the national metacomputer has no vendor who takes the responsibility for balancing all the components. The speed of all technologies in the metacomputer must increase in harmony, or bottlenecks will develop. The coordination and cooperation among vendors, national centers, and the research community needed to achieve this goal is truly a grand challenge in its own right.

None of these incredible improvements in hardware can bear fruit without corresponding increases in the capabilities of the application software. The story of how software advances turn supercomputers into tools for scientific discovery and engineering design is the subject of the next chapter.

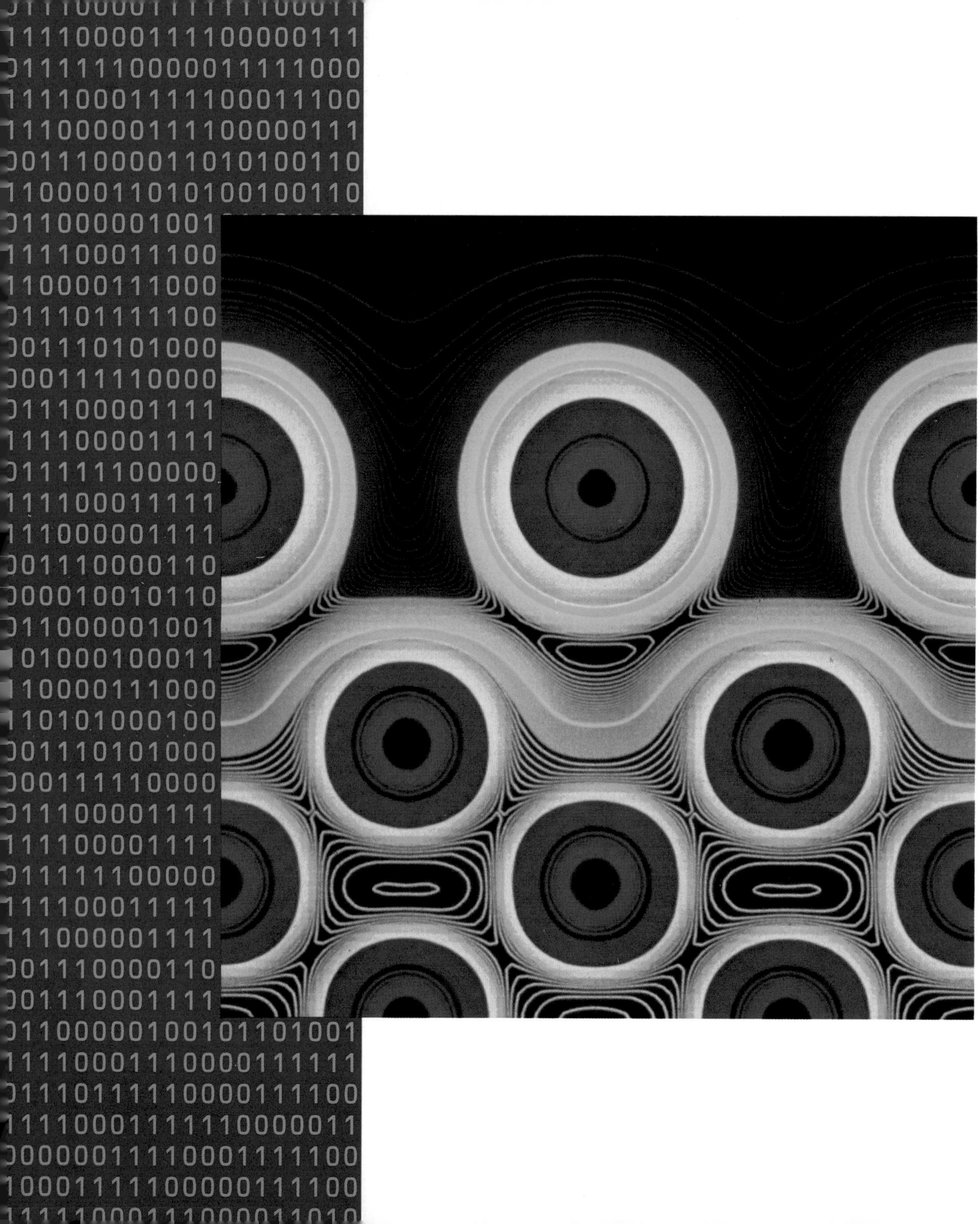

The Methods of Supercomputing

The secret of all simulation methods lies in the details of how we move from a continuous to a discrete model of the world, thereby enabling us to use digital computers to solve the laws of nature. As we shall see, the character of the laws varies greatly when we move between the classical world of macroscopic objects and the quantum mechanical world of atomic particles. As a consequence, researchers in the two realms use fundamentally different approaches to creating digital realities. In this chapter we will investigate the major classes of approximation methods that are used within the classical and quantum worlds. We will try to capture the essence of these methods and then describe their application to specific problems to illustrate the styles and strategies for using them. The problems are more idealized than those we will study in later chapters so that we can focus on the approximations themselves.

This image from work by Freeman and Wimmer shows the charge distribution in a tungsten slab with a cesium overlayer. The cesium atoms are along the top and the tungsten atoms are below. Charge density is portrayed according to colors of the rainbow, ranging from red for a high density to purple for low.

The direct logic of the computer would seem most easily applied to the phenomena of classical mechanics. These are phenomena that operate on scales large enough for us to neglect the probabilistic effects of quantum mechanics. The phenomena of the classical world range from the motion of atoms in large molecules to the behavior of materials under stress to the complex motions of fluids or gases. The equations that govern these systems are based on Newton's laws of motion and on the laws of conservation of mass, energy, and momentum, together with laws governing the forces of gravity and electromagnetism. These laws may be applied to calculate each individual particle's position and velocity, or they may be applied to calculate averages over large numbers of particles, as in the case of a gas or liquid, for which one computes the rates at which the density, velocity, and temperature of the fluid change over time.

The laws of classical mechanics describe a deterministic, clockworklike world: if you know a particle's position and momentum, and if you know all the forces acting on it, you can predict the particle's subsequent behavior fully and precisely by solving the relevant equations of motion. This straightforward approach is not possible in the submicroscopic world of atoms and subatomic particles, because quantum effects introduce a degree of uncertainty in the values of physical variables. For example, according to Werner Heisenberg's famous uncertainty principle, it is impossible to know simultaneously both the position and momentum of a subatomic particle with absolute precision. The best one can do is calculate probabilities for these and other physical quantities. To do so, one must solve an all-important equation formulated in the 1920s by Edwin Schrödinger that incorporates the tenets of quantum mechanics. The Schrödinger equation is the fundamental, governing equation of the quantum world, just as Newton's laws of motion are the principal rules that describe the classical world.

We will see how scientists re-create a probabilistic world in the computer after first describing the three major families of methods for discretizing classical problems: finite differencing, the finite element method, and particle methods.

Finite Differencing

In exploring the behavior of natural phenomena, scientists are usually interested in how these phenomena change over time. We can imagine natural phenomena embedded in a spacetime continuum divided into an infinite number of events infinitesimally close to each other, each event representing one location in space and one moment in time. As we move from the continuous world of nature to a discretized world that the supercomputer can handle, we replace the continuum of space and time with a grid of points. This grid is comparable to the grids we saw in Chapter 1, with the addition of an extra axis for the time dimension. The actual geometry of that grid depends on the particular problem that a researcher is investigating.

Many problems allow scientists to set up a regular grid of squares or boxes to approximate the spacetime continuum. These are the problems that may be addressed by the finite difference method. A simple example of a grid involving one dimension of space and the single dimension of time is shown in the drawing on the facing page. The distance between points in the space direction is a set value called Δx, and the distance between points in the time direction is a set value called Δt. As shown in the magnifying glass, the index j labels columns in the space direction and the index n labels rows in the time direction. A continuous variable, such as the temperature T that is a solution to a heat diffusion equation, can be replaced by a finite array of values labeled $T(j, n)$ to indicate the value of T at the jth point in space and the nth

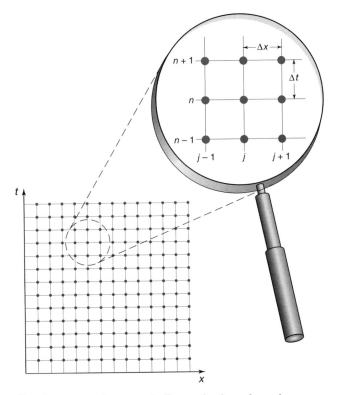

To solve an equation numerically, a scientist replaces the continuum of spacetime with a mesh or grid of points labeled by integers j for space and n for time. The points are separated by a spatial distance Δx and a time interval Δt.

number of grid points. These algebraic equations are easy to solve, and by solving them over a fine enough grid, we can make the resulting numerical solution approach a solution to the original, exact equations.

To solve a finite-differenced equation, we must begin with some specific information about the values of relevant physical variables at an initial moment of time, usually designated $t = 0$. Such information constitutes the initial conditions. These conditions may include the results of physical measurements, such as the radar data recorded at the beginning of a thunderstorm in Wilhelmson's storm simulation, or they may derive from a mathematical statement, like the small perturbation introduced by Woodward into the smooth interface between two fluids. This information serves as a springboard from which we can jump forward in time in steps of Δt. FORTRAN DO loops are well suited to describe the repetitive steps needed to solve the same equation over and over, moving from one grid point to the next, thereby weaving out the fabric of the solution. In order to specify how conditions outside the domain of our calculations interact with the region we are computing, we need to pose boundary conditions at the ends of the spatial grid. In Chapter 1, we saw that Woodward used periodic boundary conditions, which stated that the values at each end of the grid were identical, while Wilhelmson used boundary conditions that let material flow freely off the grid.

The flow of heat down a metal rod provides an example of how one uses finite differences to solve a problem. Imagine a metal rod with one end embedded in a block of ice while the other end is heated with a blowtorch. Initially, the temperature along the bar increases linearly from 0°C at the cold end to some high temperature (T_0) at the hot end. We then turn off the blowtorch and ask how the temperature along the bar changes with time.

The way in which heat diffuses or spreads throughout a substance is described by a diffu-

moment in time. As the program for the simulation runs, all the values would be calculated one row at a time, for the desired number of time intervals, until the last row is complete.

Like all supercomputing methods, the method of finite differencing calls for the relevant equations to be transformed into a form that allows them to be solved numerically. To do so, the rates of change, or the "derivatives," in the equations are replaced with algebraic expressions involving the finite quantities Δx and Δt. The original equations, which are defined at every point in space and time, thus become algebraic equations accurate only at the finite

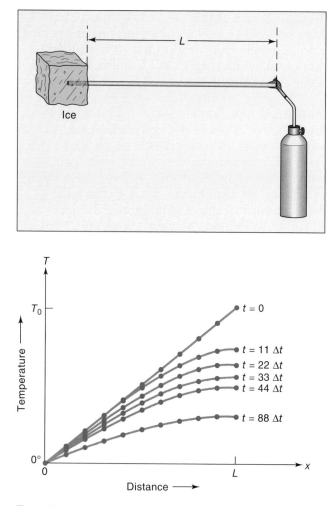

Top: *Finite differencing can be used to predict how the temperature along the metal bar changes after the blowtorch is shut off.* Bottom: *The dots on the graph are numerical solutions that predict the temperature at 10 locations along the bar at selected times. For comparison, the curves are exact solutions.*

sion equation. According to this equation, the evolution of temperature at every point along the metal bar depends on how much the temperature differs between adjacent locations, adjusted for the thermal conductivity of the bar and its capacity to store heat. The equation gives precise expression to our intuitive notion that heat flows from the hot end of a metal bar to the cool end.

To set up the grid, suppose we divide the bar into 10 segments so that $\Delta x = L/10$, where L is the length of the bar. Thus the index j runs from 0 at the cold end of the bar to 10 at the hot end. The n index starts at zero, corresponding to the initial moment $t = 0$ when we turn off the blowtorch.

For our initial conditions, we shall assume that the temperature at time $t = 0$ varies uniformly from 0°C to T_0 at all 10 points along the bar. Inserting these initial conditions into the finite-differenced equation, we compute the temperature at all 10 points at time $t = \Delta t$. This new data is fed back into the finite-differenced equation so that we can calculate the temperature along the bar at the next moment of time, $t = 2\Delta t$. In terms of a spacetime diagram like that on the previous page, data from the nth row inserted into the finite-differenced equations gives the values of the temperature along the $(n + 1)^{\text{th}}$ row. In this repetitive way, we move forward into the future as far as we choose in steps of Δt.

The dots in the graph on this page indicate numerical solutions to the finite-differenced equations. The curves give exact solutions, which happen to be available for this comparatively simple example. As you would intuitively expect, the solutions predict that the bar of metal simply cools off as heat is drained away by the block of ice.

The resulting numerical solution solves the finite-differenced equation, not the exact heat diffusion equation. In general, the finer the grid, the closer the numerical solution is to the exact solution of the laws of nature. However, unless one chooses the algorithm for solving the discretized equations carefully, finite differencing can be subject to difficulties called instabilities that can cause a numerical solution to gyrate wildly.

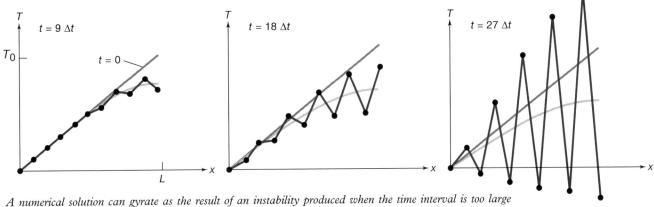

A numerical solution can gyrate as the result of an instability produced when the time interval is too large compared to the space interval. The amplitude of the characteristic sawtooth pattern grows with time.

Suppose you were tempted to speed up the calculations by taking larger time steps. The graphs above show that when Δt is chosen too large, the numerical solution begins to deviate from the exact solution early in the computations and that these deviations grow dramatically in the next few time steps. Such solutions are said to be unstable. The problem can be corrected by keeping Δt small enough or by choosing a more complex algorithm that allows Δt to remain larger, but this makes the program longer to execute.

Even if the chosen algorithm is stable, there are still two fundamental sources of error that can cause a numerical solution to be inaccurate. The first is the truncation error, which refers to the accuracy with which a few algebraic terms can approximate a rate of change expressed exactly in the original equation by a derivative. Finite differencing methods replace a derivative by an infinite series of terms in powers of Δx. One then cuts off or "truncates" the infinite series, leaving only a few terms. The loss of terms may lower the accuracy of results, introducing the error. To reduce the truncation error, we must either use a finer grid (smaller Δx) or include more algebraic terms in the expressions that replace derivatives, which results in an algo-

rithm of "higher order" accuracy. Either strategy requires taking more execution time or turning to faster, more powerful supercomputers.

The second source of error is roundoff, which refers to the dropping of digits. In any calculation, the numbers that the computer manipulates must have a finite length (most supercomputers carry up to 13 decimal digits per variable), and so the last few digits in a long number may be discarded. Roundoff error is easily detected by making some small changes in the problem, such as a small adjustment in the initial conditions. If such tiny changes result in large modifications in the solution, then roundoff error is the likely culprit. One goal of algorithm design is to create approximation schemes that are insensitive to roundoff error. Errors caused by truncation are usually orders of magnitude larger than those induced by roundoff errors.

The sometimes subtle effects of instabilities, truncation errors, and roundoff errors may not be readily apparent. Scientists do not trust the accuracy of their results simply because no flaws are obvious. For every simulation, they follow a painstaking set of procedures designed to reveal distortions and errors. We can trace their steps by examining the finite differencing of supersonic gas jets.

Testing a Simulation: Supersonic Gas Jets

The mix of hot gases shooting from the rear of a rocket or a supersonic jet may resemble a chaotic swirl to the casual observer, but a closer look reveals definite structures within the flow. Sometimes these features are obvious, as you can see in the photograph of the SR-71 Blackbird jet taken in 1990 during its final record-setting flight across the United States. Note that the supersonic exhaust is quite straight and contains repeated bright knots.

When gases are moving faster than the speed of sound, we can expect shocks to dominate the flow behavior, as we saw in the Woodward simulation in Chapter 1. The Blackbird's exhaust is emerging into the atmosphere faster than the speed of sound, creating weak repeated internal shocks that cause alternating regions of compression and rarefaction along the gas jet. The bright knots appear in areas of compression where the internal shocks in the jet are re-igniting unburnt fuel.

The launch of the NASA Space Shuttle produces a more dramatic shock in the exhaust. Here one can see embedded in the supersonic flow a bright disk, which is a flat shock termed

a Mach disk. Downstream from the disk-shaped shock are more repeated internal shocks. These two distinct types of shocks, weak repeated internal shocks and strong transverse shocks that disrupt the flow, are fundamental features of supersonic jets on cosmic scales as well.

Jets millions of light-years in length emerge from the centers of certain galaxies. One of the strongest sources of radio emission in the sky, Cygnus A, is a double jet ending in two radio-emitting lobes positioned rather symmetrically on either side of an active galaxy. As we shall see in Chapter 8, such extragalactic jets are believed to be the result of gas accreting onto a supermassive black hole at the galaxy's center. Bright knots of radio emission develop along each of the two jets as it moves outward from the galactic core. The jets are straight for a long distance, then begin to wiggle. Both end in a large, bright plume of radio emission having a "hot spot" of emission embedded at its end.

The challenge to physicists, who believe that their mathematical laws explain all scales of gaseous phenomena, is to try to understand these different phenomena over an enormous range of lengths and times by a single computational approach. That challenge was taken up in Munich in 1981 by one of the authors, Larry Smarr, and his colleagues Michael Norman and

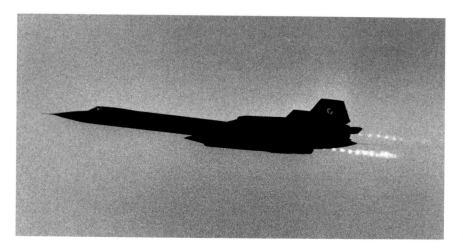

Regularly spaced bright knots illuminate the supersonic jet exhaust of the SR-71.

The circular base of the blue "cone" seen embedded in the rocket exhaust of the Space Shuttle orbiter's engines (right) is a Mach disk standing shock.

Karl-Heinz Winkler, using the Cray-1 at the Max Planck Institute for Physics and Astrophysics. Since then, work has continued, including some of that described below, at Los Alamos National Laboratory and at the National Center for Supercomputing Applications (NCSA).

Although the physical situations in the various examples above differed, we felt that the basic structures of the jets would be revealed by studying solutions of the simplest gas dynamics equations, the three Euler equations. The first equation expresses the conservation of mass, keeping track of how the density changes in any grid zone as a result of matter flowing into or out of that zone. The second and third equations express the conservation of momentum and energy. Finally, a fourth equation, called an equation of state, relates gas pressure to density and energy, thereby characterizing these physical properties of the gas. We have already encountered in Chapter 1 two examples involving the Euler equations, Woodward's shear flow and Wilhelmson's severe storm.

The Euler equations realistically treat gas as compressible, but ignore any viscosity, magnetic fields, charged particles, or combustion

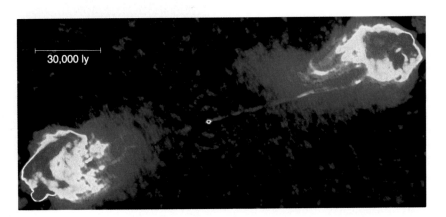

Two jets stream outward from the core of a galaxy, forming one of the brightest radio objects in the sky, the double lobe radio source called Cygnus A. The color is proportional to the radio emission in this radio map made by R. Perley and J. W. Dreher with the Very Large Array radiotelescope.

that influences the jets in the real world. Eliminating these factors ensured that the gas would behave the same regardless of the jet's physical size, and thereby improved our chances of successfully simulating a phenomenon whose length can range from a few meters to millions of light-years. This process of abstracting the simplest mathematical representation that will capture the major features of a complex physical situation is a recurring theme in computational science and engineering.

After first simplifying the physics, we next simplified the geometry, to make the problem tractable for the supercomputers of the early and mid-1980s. We assumed the gas flow to be symmetrical about an axis, because real-world examples of supersonic jets are dominated by that symmetry. The assumption of axial symmetry allowed us to tackle the dynamics in two dimensions rather than three: we could simply divide a two-dimensional plane into squares or rectangles and then imagine rotating the solution around the axis of symmetry to obtain the full three-dimensional solution. Indeed, because of the rotational symmetry, we needed to compute only the upper half plane of the gas flow.

Since real-world jets propagate into a pre-existing atmosphere, we filled the grid at the initial moment with a uniformly distributed gas at rest. The jet gas enters the grid through an orifice in the middle of the left wall and bores its way through the ambient gas that fills the grid until the computations are halted just as the advancing jet is about to strike the right wall. As the ambient gas is displaced by the advancing jet, it is allowed to escape through the outer boundaries of the grid. We assumed that the gas pressure in the jet is equal to the pressure exerted by the ambient gas. The jet is therefore said to be pressure-matched, which seems to be roughly the case in astronomical jets.

It is convenient to characterize pressure-matched jets with two quantities. First is the jet's Mach number (M), the ratio of the jet speed to the speed of sound in the jet gas. Second is the density ratio η, which equals the den-

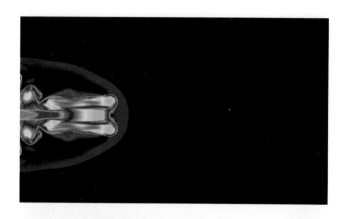

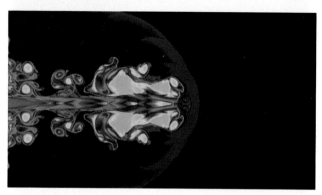

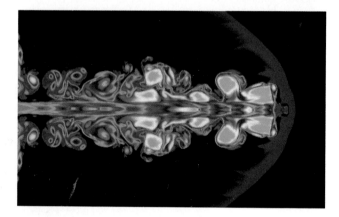

A supersonic jet evolves in form as it penetrates an ambient, stationary medium. The speed of the incoming gas in this standard model is six times the speed of sound and its density is a tenth that of the ambient gas. The impact of the injected gas on the ambient gas produces a bow shock (purple) that propagates ahead of the beam. Colors display the logarithm of the density.

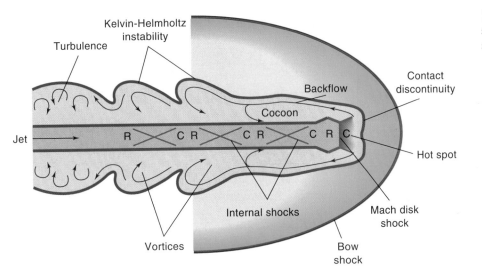

This schematic drawing illustrates the main features of a mature, standard model Mach jet.

Turbulence
Kelvin-Helmholtz instability
Backflow
Contact discontinuity
Cocoon
Jet
R   C R   C R   C R C
Hot spot
Internal shocks
Mach disk shock
Vortices
Bow shock

sity of the jet gas divided by the density of the ambient gas. In the three views shown on the opposite page from a simulation by Michael Norman, a supersonic jet propagates with a Mach number of 6 and a density ratio of 0.1. Thus the ambient gas is 10 times denser than the jet gas, which enters through the orifice at the left of each view at six times the speed of sound. Hereafter we will refer to these physical parameters as providing the standard model of the jet.

By referring to the schematic diagram above that identifies the main features of the standard model, we see that the rightward propagating jet is the straight, narrow feature extending across the middle of each image. Careful examination of this slender feature reveals that as the jet travels from left to right, it periodically expands and converges in a series of internal shocks, similar to those seen in the exhausts of both the Blackbird and the Space Shuttle. The supersonic flow terminates at the head of the advancing beam in a feature called the working surface. This shock is of exactly the same type as the Mach disk in the Space Shuttle's exhaust, and this type of shock is also the cause of the "hot spots" of radio emission seen in Cygnus A. This strong transverse shock abruptly brings the

jet flow from supersonic to subsonic speeds, severely heating it, and deflects it sideways out of the path of the beam. The gas then flows backward toward the orifice, forming an enveloping "cocoon" filled with turbulent eddies. This cocoon of shocked gas corresponds to the plume surrounding the jet and hot spot in Cygnus A. Finally, a bow shock, which precedes any supersonic object such as a high-speed bullet or the Concorde, surrounds the cocoon.

To completely understand any solution to a set of equations, it is necessary to look at all the variables that define the solution. Rather than mapping density for the standard model, the left image on the next page maps the divergence of the velocity field, a measure of how abruptly the speed or direction of the gas flow changes from one location to another. High values of the divergence reveal the location of shocks, because shocks are characterized by an abrupt slowing of the gas. Because the jet is pressure-matched to the ambient gas through which it is boring, the boundaries between the beam and the cocoon and between the cocoon and the ambient gas are invisible in the image on the right mapping the logarithm of the pressure variable, even though density jumps a hundred-fold across the boundary separating the cocoon from the ambi-

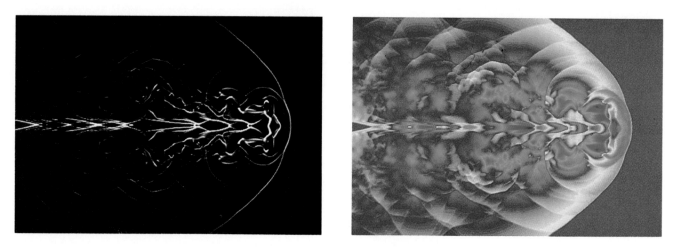

Two views of the axisymmetric, pressure-matched standard model jet each mapping a different physical variable: divergence of the velocity field (left) and the logarithm of the pressure (right).

ent gas. On the other hand, the jump in pressure across the internal and bow shocks is greater than the jump in density, so these shocks are more distinct in the mapping of pressure.

How do we know that we can believe these detailed substructures in the finite difference simulations? Computational scientists use three fundamental checks to validate their methods. First, they run a series of simulations of the same physical situation, but reduce the spacing between the grid points for each run. As the resolution rises, the simulation should converge to one solution. Second, they rewrite the code using different numerical algorithms to see if the same solution appears independently of the numerical methods.

We carried out both tests using our jet code on the standard model. The upper half of the left image on the facing page shows a standard model jet computed on a grid of $400 \times 600 = 240,000$ zones, each zone a rectangle twice as long in the direction of the jet as across the jet. At this resolution, the radius of the rightward-moving jet was covered by 20 zones. That single half image took more than 20 hours to complete on the NCSA Cray Y-MP.

The algorithm used to solve the gas dynamics is termed "second order accurate" to indicate that the truncation errors are proportional to $(\Delta x)^2$. If we run the code with the same algorithm, but one-quarter the number of zones in each direction for a total of 15,000 zones, then the resolution of the solution is much worse, but the computation takes only 20 minutes.

The truncation error is the amount by which the discretized equations are less accurate than the exact equations. Since Δx is assumed to be very small, higher powers of Δx make the truncation error smaller and smaller and thus make the solution more and more accurate. Rewriting the algorithm to achieve higher order accuracy is an alternative route to obtaining more realistic simulations.

If we use the coarse grid and a simpler algorithm, say one in which the truncation error is proportional to Δx, then the solution captures none of the jet's features except for the bow shock (right image). The simpler algorithm would be easier to program, but the running time of its code is practically the same as that of the second order accurate methods. Keeping the coarse resolution, but going to a more complex, third order accurate method restores many of the general features of the finely zoned run, but the computing time is doubled by the complexity of the numerical operations. Even when used with the higher order method, the coarse grid does not allow the simulation to capture the breakup of the cocoon. Since the motions of the cocoon perturb the jet, additional shocks develop in the jet at higher resolution. These internal shocks do not disrupt the jet propagation, but they do slow it down, which is why the length of the jet becomes shorter at higher resolution.

The lesson is that we can improve the accuracy of a simulation either by refining the zoning or by using more complex algorithms. Either of these options increases the supercomputing time needed, unless we can acquire a faster supercomputer. However, a more complex algorithm may be able to achieve a more accurate solution for less increase in supercomputing time than will adding finer grid zones. The more complicated equations in a rewritten algorithm may take three or four times longer to solve, but refining the grid increases the computing time by powers of the grid refinement factor. Nonetheless, you cannot model features such as cocoon vortices if they are smaller than the grid zones no matter how good the algorithm. The desire of scientists to achieve the high accuracy that can ultimately only be supplied by using computationally demanding finer grids explains why they are always demanding faster supercomputers with larger memories.

The code for modeling a supersonic jet passes the first two tests, since as both grid refinement and algorithm improvement increase the physical resolution, our solution "comes into focus." Once the consistency of the code has been verified, the scientist checks that the code can reproduce known experimental results or theoretical expectations. As we saw above, our code predicts shock wave structures in the jet of just the sort seen in the supersonic exhausts of the Blackbird and the Space Shuttle. It also reproduces the overall behavior seen in radio maps of extragalactic jets.

Next, the scientist needs to decide whether the general features of the solutions continue to

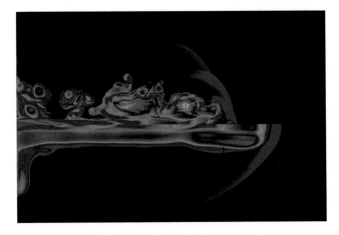 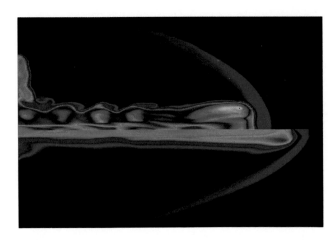

Left: *This split view of the standard model jet compares a simulation run with fine zoning* (top) *to one run with coarse zoning. Both simulations were run with the same second order accurate algorithm.* Right: *Another split view compares a third order accurate method* (top) *and a first order accurate method* (bottom), *both run with the same coarse zoning as in the bottom half of the previous image.*

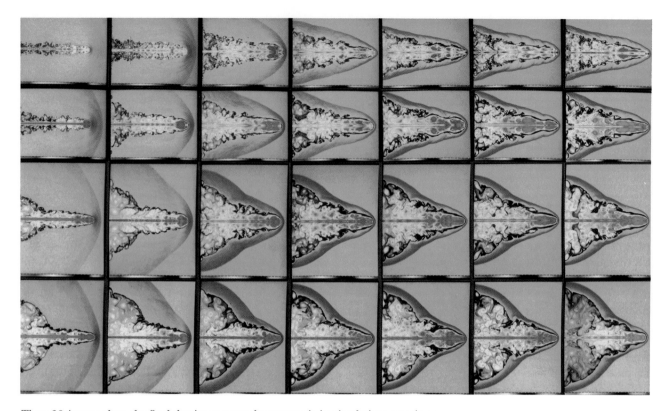

These 28 images show the final density contours for supersonic jet simulations covering a large range of parameters. From left to right, successive columns have Mach numbers (M) of 1.5, 3, 6, 12, 100, 1000, and 10,000. Vertically, from top to bottom, successive rows have density ratios (η) of 10^{-2}, 10^{-3}, 10^{-4}, and 10^{-5}. The simulation of the low Mach number, higher density ratio in the top left required only 30 hours of machine time on a Cray X-MP, but the simulation at high Mach number, low density ratio at the bottom right of the picture required 400 hours.

appear even when the defining parameters have been changed. In the case of the supersonic jets, the Mach number M and the density ratio η are the two parameters that characterize the problem. Karl-Heinz Winkler used Cray supercomputers at the Los Alamos National Laboratory to run several thousand hours of high-resolution jet runs that varied M and η by several powers of 10. The images above show the results of over two dozen numerical simulations of super-

sonic jets for density ratios ranging from 0.01 to 0.00001 and for Mach numbers from 1.5 to 1000. As you can see, the same general structure remains evident throughout: internal shocks are interspersed along a central beam, which terminates in a transverse shock that creates a backflowing cocoon full of vortices, and the whole assemblage is surrounded by a bow shock. Nevertheless, the changing parameters do gradually alter the shapes and positions of these

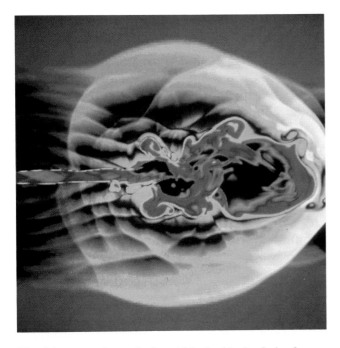

The slab symmetric standard model jet in this simulation has been perturbed and is wiggling into a large-amplitude instability. The oscillations along the jet were forbidden by symmetry in the axisymmetric case shown earlier. A particularly informative scheme for associating colors with density was designed with the help of computer artist Donna Cox.

features. Two important trends can be seen: the angle of the bow shock depends on M but not on η, and the cocoon size increases as η decreases for fixed M.

One final test is to verify that the original assumption of axisymmetry did not overly distort the results. Scientists often do not have the machine power to model the full phenomenon of interest, and they need a way to tell which features in their simplified model correspond to the real phenomenon. The test in this case is to change the geometry of the code and see which general features are still maintained. A different geometry, also in two spatial dimensions, is called slab-symmetric. Here the orifice in the left vertical wall is a long slit instead of a circular hole. The simulation by Michael Norman

shown above has the same parameters as the standard model and illustrates what a supersonic "slab jet" looks like. We again see a large cocoon and the presence of internal shocks, although they are now tilted slightly, in response to the new geometry. Phenomena that are robust enough to appear in different symmetries are likely to be phenomena that will appear in the full three-dimensional case. Two new features are the wiggling of the jet and the asymmetry of the cocoon, both visible to a lesser degree in the radio jet in Cygnus A. Phenomena that appear in one symmetry but not in the other may appear in an intermediate form in the real object.

Characteristics of both the axisymmetric and the slab-symmetric jet should be seen in a fully three-dimensional jet, if one had a supercomputer with enough memory to hold the data. The advent of supercomputers with more than one gigabyte of memory has allowed for just this development. Just as this book was going to press, Norman and James Stone ran a 40 million zone simulation of a fully three-dimensional jet with the same parameters as the standard model. This single run required over 300 hours on NCSA's new Convex C3 gallium arsenide

This visualization of a 40 million zone simulation of the standard model shows the nonaxisymmetric instabilities in the cocoon that lead to filamentary structure.

supercomputer and used over one gigabyte of memory. The analysis of this simulation, probably the most finely zoned of any gas dynamics simulation in this book, is still underway. Preliminary results show that this highly resolved three-dimensional jet far better captures the filamentary plume–like appearance seen in Cygnus A.

This example of finite differencing illustrates the variety of strategies used to establish the validity of results, such as refining the grid, changing algorithms, making comparisons with observed phenomena, covering parameter space, and changing spatial symmetries. Several of these same methods, especially covering parameter space and changing spatial symmetries, also serve as useful tools for investigating a phenomenon in greater detail, as does comparing different variables. Similar methods were used to validate and explore most of the examples in this book. The reader should therefore realize that each image in the book represents but a tiny portion of the actual number of computations executed by the researchers to obtain their final answers.

Finite Elements

Although finite differencing is straightforward, it is not always the best method of tackling the equations that govern the world around us. Because of its reliance on regular coordinate systems, finite differencing has distinct disadvantages when used to solve problems with complicated shapes. An alternative technique, called the finite element method, overcomes such difficulties by dividing a complex domain into a mesh of geometrically simple subdomains called finite elements. Like the finite differencing method, the finite element method transforms the original equations, expressed in terms of derivatives, into algebraic equations. These transformed equations, called element equations, can be applied to each finite element separately. Because of the small size and simple shape of each finite element, the derivation of the re-

quired element equations is usually not terribly difficult. Terms in the element equations are numerically evaluated for each element, and then the values obtained are inserted into a much larger set of algebraic equations, called system equations, which are solved for the problem at hand.

In many respects, the finite element method is the inverse of finite differencing. Finite differencing uses a global, regular coordinate system to divide a computational domain into small segments, whereas the finite element method views each small segment as a basic unit and connects them together to assemble the computational domain. Thus finite element methods are very useful for irregularly shaped domains, for domains requiring a mix of different shaped elements, or for domains in which certain critical areas require special meshes for accuracy (as we shall see in the crack propagation example below). Many engineering problems have these characteristics, and finite element methods are frequently employed in these problems.

The basic ideas underlying the finite element method can be illustrated in determining the area of an ellipse. First, the computational domain (i.e., the ellipse) is represented by a collection of a finite number of subdomains, say rectangles. Each rectangle is called an element, and the collection of elements is called the finite element mesh. Two alternative rectangular meshes are shown on the facing page: "Mesh A" circumscribes the ellipse, whereas "Mesh B" is inscribed within the ellipse. We then compute the desired property for each isolated element, in this case the area. It is here that we introduce the governing equation to obtain the element equations: an element's area equals its base times its height. The approximate area of the ellipse is then obtained by assembling the solutions to the element equations. In this example, the total area simply equals the sum of the areas of the individual elements.

As you might expect, the finer the mesh, the closer the computed area is to the actual

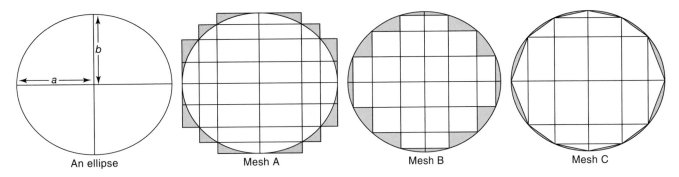

An ellipse   Mesh A   Mesh B   Mesh C

Any of these three meshes can be used to compute the area of an ellipse, which equals
πab, *where* a *and* b *are the semi-major and semi-minor axes of the ellipse, respectively.*
The computed areas converge on the true value as the number of finite elements is
increased.

area of an ellipse, which equals πab, where *a* and *b* are the semi-major and semi-minor axes of the ellipse, respectively. The graph on this page shows how the computed area converges on the true value. Note that a combination of rectangles and triangles covers the ellipse better than rectangles alone, and so the area computed using "Mesh C" converges even faster.

The use of mixed shapes as elements is quite common in finite element problems. When

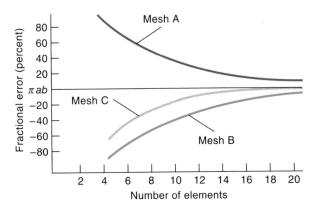

This graph shows how the areas computed using the three meshes at the top of this page converge to the area of an ellipse (πab) as the number of elements increases. The precise shapes of the curves in this graph depend on how the various elements are drawn.

designing a mesh for a particular application, the scientist or engineer intentionally makes the mesh elements as simple as possible. Shapes such as triangles and rectangles are common in two-dimensional problems, while "bricks," "wedges," and "pyramids" are used to cover three-dimensional domains. During the early years of finite element analysis, researchers created meshes by hand, usually on large sheets of graph paper. This tedious, error-prone procedure has been replaced by computer programs that automate the process of mesh generation. For three-dimensional simulations especially, generating a complex mesh is so numerically intensive that it may require a supercomputer.

As an example, one finite element mesh used to calculate details of air flow past a complete aircraft consists of 720,859 tetrahedra (i.e., pyramids) that fill a cylindrical volume completely surrounding the aircraft. The mesh was generated using a program created by J. Peraire, J. Peiro, and K. Morgan in the Department of Aeronautics at Imperial College, London. The program began by covering the airplane with 30,628 triangular elements, a process called triangulation of the computational boundary. These triangles formed the bases of tetrahedral elements that projected from the surface of the aircraft. The faces of these tetrahedra constituted the bases of yet more tetrahedra, as the computer moved outward from the aircraft filling the entire computational domain with tiny, tightly packed pyramids. The program that produced this mesh generated tetrahedra at the rate of 300,000 per hour on one processor of a Cray-2. Using this tetrahedral mesh, Peraire, Peiro, and Morgan were able to compute the distribution of pressure on the aircraft's surface.

Gaseous or fluid flow, which had traditionally been the domain of finite differencing, is one of the fastest growing areas for finite elements in three dimensions. Many other areas of engineering, science, and applied mathematics are also well suited to finite element methods. Applications have occurred principally in solid mechanics (statics, vibrations, dynamics, large deformations, fracture, plasticity, and elasticity); heat transfer by means of convection, conduction, and radiation; electromagnetism; and acoustics. New areas of application, such as solid state physics and quantum mechanics, are being explored and developed. A case from the mechanics of solid body fractures illustrates how the flexibility of finite element methods can be used to achieve higher resolution.

Cracks are present in essentially all structural materials, either as natural imperfections or as the result of manufacturing processes. Moreover, different materials fracture in different ways. For instance, the metal in bridges and airplanes is subjected to repeated loads and can exhibit so-called fatigue fractures, which propagate relatively slowly. Hence regular inspection can find a small crack before it becomes large enough to be dangerous. However, substances like aluminum or metals at very low temperatures exhibit brittle behavior. When subjected to an excessive load, these materials fail suddenly and fracture without warning. If engineers understood precisely what happens when a crack moves through a material, they would be better able to predict when cracks might start, how fast they would grow, and under what circumstances they might stop growing.

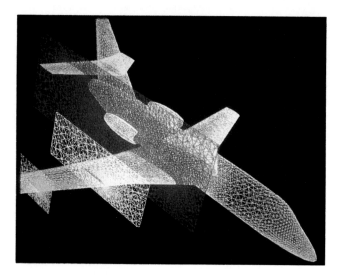

A Cray-2 generated this tetrahedral mesh to cover the computational domain surrounding an aircraft. The image shows the arrangement of the mesh on the airplane's surface and on three planes that cut through the domain.

At the University of Illinois at Urbana-Champaign, Robert Haber, Hyun Koh, and Hae-Sung Lee used finite element methods on the Cray X-MP to examine how cracks, especially those that grow at very high speeds, move through brittle substances that fracture suddenly when subjected to excessive loads. An innovative feature of their work is the way in which they set up the mesh over which the required laws of solid body mechanics are expressed in finite element form. In most solid mechanics problems, engineers use a Lagrangian frame: the mesh is tied at every point to the material under investigation, like lines etched onto a steel plate. In fluid mechanics problems, such as a simulation of a supersonic jet, researchers generally use an Eulerian frame: the mesh is fixed in space, and material flows across it.

Haber, Koh, and Lee decided to use a combination of these two extremes, a mixed

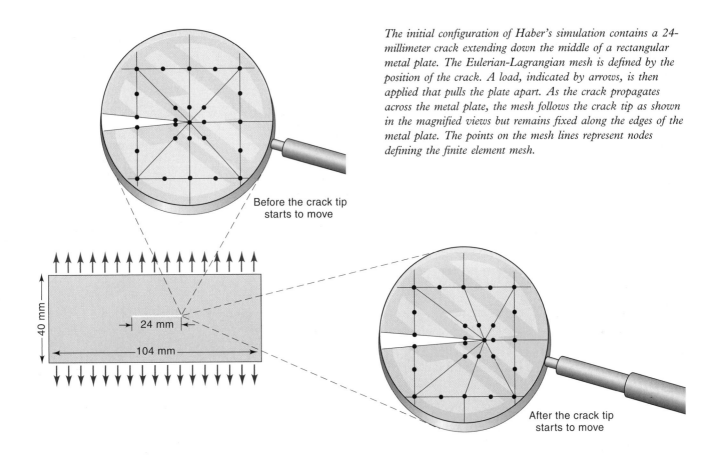

The initial configuration of Haber's simulation contains a 24-millimeter crack extending down the middle of a rectangular metal plate. The Eulerian-Lagrangian mesh is defined by the position of the crack. A load, indicated by arrows, is then applied that pulls the plate apart. As the crack propagates across the metal plate, the mesh follows the crack tip as shown in the magnified views but remains fixed along the edges of the metal plate. The points on the mesh lines represent nodes defining the finite element mesh.

Before the crack tip
starts to move

40 mm

24 mm

104 mm

After the crack tip
starts to move

Eulerian-Lagrangian scheme, because the material particles located at the crack tip are constantly changing as the crack tip moves. Their mesh is anchored around the boundaries of the objects they investigated, but near the center the finite elements are tied to the moving crack tip, so that the mesh changes shape as the crack develops. As a result, there is always a finer grid to help resolve the rapidly moving crack tip.

In their simulation, Haber and his colleagues examined the behavior of a rectangular plate with a central crack when a load is applied to both sides of the plate. About an hour of computing time on a Cray X-MP was sufficient to calculate a variety of physical quantities. For example, the supercomputer evaluated the so-

called stress intensity factor as the plate was pulled apart; this quantity is a measure of the force per unit area, for a specific crack length, acting at a particular location. The strain energy density, which is related to the deformation of the plate, was also calculated. Another computed quantity, the kinetic energy density, is proportional to the square of the material velocity and thus reveals how fast various regions of the plate are moving.

The finite element simulation reveals the logical sequence of events after the plate is pulled apart. The load applied at the edges of the plate causes a stress wave to move inward parallel to the sides of the plate until it hits the crack 4.4 microseconds later. The action of the

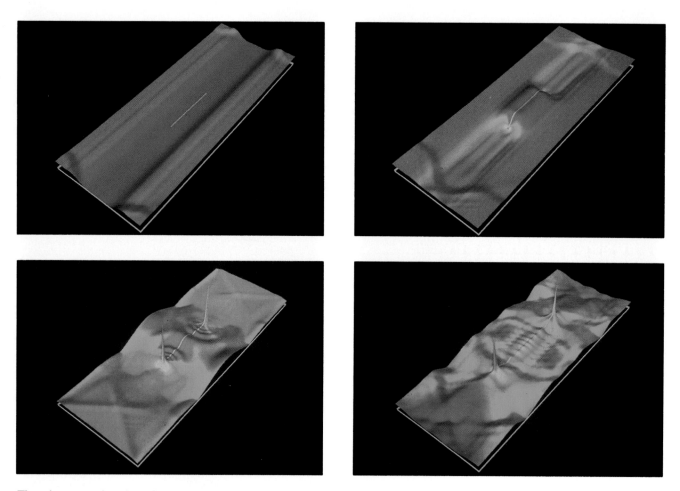

These four successive views show how two important physical quantities change as a central crack propagates down the middle of a rectangular metal plate. The vertical relief, proportional to the kinetic energy density, is related to the speed with which the crack and associated waves move, whereas color represents the strain energy density, a measure of the deformation of the plate. The color map uses blue for low density and red for high, with intermediate values given by intermediate colors of the rainbow. The elapsed time for the four snapshots ranges from 0 to 15 microseconds.

load also creates diagonal stress waves that move inward from the corners of the plate. When the stress wave arrives at the plate's center, both tips of the crack begin to move at the high velocity of 1.0 kilometer per second. Like the ripples created by a stone thrown into a pond, beautiful circular elastic waves begin to radiate outward through the material as the crack tip

travels. When these waves reach the edges of the plate, they are reflected back and return to the crack tip to interact with the stress field.

Previous work on crack propagation had plotted quantities like the stress intensity at the crack tip as a function of time. Indeed, an exact solution for that quantity exists for the case of an infinite plate. Haber's simulation was able to

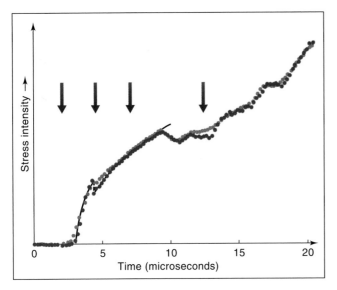

A graph of the stress at the end of the crack tip versus time compares an exact solution that assumes the plate is infinite in area (the solid line) with two different finite element analyses of the finite plate problem (the dots and squares). The agreement between the exact and numerical solutions at early times verifies the accuracy of the code, but only the code can give the complex behavior at later times. The arrows indicate the four moments during the evolution depicted in the images on the opposite page.

explain a divergence between the exact solution and numerical solutions that had already been known to exist: the finite element solution begins to deviate from the exact solution at the moment the reflected waves reach the crack tip because an infinite plate has no edges to produce reflected waves. Subsequent peaks and valleys in the plot of the stress intensity factor are associated with the arrival of elastic waves at the crack tip. By showing the stress intensity at all points on the plate rather than just at the crack tip, the images from the simulation clearly reveal the sequence of propagating waves that explain the diverging analytical and numerical results, reassuring the scientists that the divergence results from the physics, not from a faulty code. The wave patterns and their effects in crack propagation have been far better detailed in

these numerical investigations than in anything available from either elaborate laboratory experiments or painstaking exact analyses.

Particle Methods

Astronomical bodies such as stars, planets, and galaxies move along in paths determined by the gravitational forces of their neighbors. The basic equations of motion describing these paths were written down three centuries ago, but exact solutions exist for only the very simplest case. Indeed, Isaac Newton's classical two-body problem, wherein two objects travel along elliptical orbits about their center of mass, is the only scenario for which exact solutions exist. The interactions of multiple bodies require numerical methods.

Newton's second law states that the acceleration experienced by a particle is equal to the force acting on that particle divided by the mass of the particle. If you know all the forces acting on the bodies of a system at any moment, Newton's second law tells you the acceleration that each particle experiences and thus you can calculate the change in each particle's velocity, which will then determine where the particles will be a short time later. In this new configuration, you then recalculate the forces on all the bodies, obtaining new accelerations with which you can compute the locations of all the bodies at the end of the next time step. In this repetitive manner, so ideally suited for a supercomputer, you march the system of bodies forward in time. Proceeding onward by this means, you can eventually obtain a solution to any so-called N-body problem.

In a system of N bodies, the process of computing the force exerted on each body by all the others requires $N \times N$ calculations. If N is a large number, N^2 computations for each time step can be prohibitive, even for a supercomputer. In the mid-1980s, A. W. Appel, followed by J. Barnes and P. Hut, D. Porter and G. Jernigan, and L. Greengard and V. Rokhlin, independently developed new methods for N-body problems. These methods are called tree codes or hierarchical N-body codes, and they can be significantly faster than classic direct-sum algorithms when applied to large calculations. The relative advantage of a tree algorithm grows as the number of particles is increased. A typical tree method may be only about 10 times faster than a direct-sum algorithm for $N = 10,000$ particles, but nearly 100 times faster for $N = 100,000$ particles, and almost 1000 times faster for $N = 1,000,000$ particles.

Tree algorithms group distant particles together and treat them collectively, insofar as their effects on a given particle are concerned. At the lowest level of the tree code are the individual particles themselves. The next layer up is constructed from groups of neighboring particles, which are replaced by hypothetical particles located at the center of gravity of each pair. Successively simpler layers are constructed in the same fashion, as each new layer is derived from its predecessor by combining groups of neighbors.

Tree codes may be less exact than a straight summing up of all forces, and are significantly more complex to program. However, where they can be used, they can make intractable N-body problems routine. Both direct and tree methods can be used successfully on massively parallel machines, because the forces between pairs of particles are computed independently of the forces on other pairs of particles. Thus the forces for different pairs can be computed simultaneously on many processors. In 1989, J. Salmon, M. Warren, and P. Quinn at Caltech demonstrated that a variant of the Barnes-Hut algorithm ran 350 times faster on a 512-processor nCUBE than on a single processor, and in 1992, Salmon and Warren ran another version of the algorithm on the Intel Delta that executed 445 times faster on 512 processors than on a single processor, for a total speed of 5.4 gigaflops.

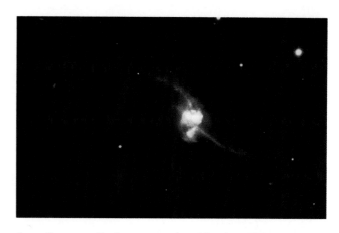

Long "antennae" of stars were ejected by the collision of a pair of galaxies called NGC 2623.

The collision of two galaxies is an N-body phenomenon that has been explored using particle methods. A galaxy consists of roughly a hundred billion stars all orbiting their common center of mass under the influence of their mutual gravitational interactions. This fact alone makes galaxies ideal for modeling by N-body techniques. But galaxies are not scattered randomly around the universe; rather, they are grouped together in clusters. In a cluster, galaxies follow trajectories that are dictated by the gravitational forces exerted on them by all the matter in the cluster. During their travels, two galaxies occasionally pass close enough to collide, yet there is so much space between the stars that the probability of two stars crashing into each other is extremely small. Nevertheless, the gravitational interaction between colliding galaxies can produce spectacular phenomena, observed by telescopes, as they hurl thousands of stars out into intergalactic space, along huge, arching streams.

During the late 1980s, Joshua Barnes, now at the Institute for Astronomy in Hawaii, began using supercomputers at the John von Neumann Center at Princeton and at the Pittsburgh Supercomputing Center to simulate galaxy collisions using models that were quite realistic.

Barnes's galaxies included the three main components commonly observed in most galaxies: a central budge densely packed with stars, a rotating disk of stars, and a large spherical halo surrounding the disk and containing scattered stars and dark nonstellar matter. An efficient N-body tree code algorithm that combined the gravitational effects of stars made it possible to run simulations with more than 30,000 particles per galaxy without imposing artificial restrictions on the shapes or extent of the galaxies.

The simulation by Barnes shown on this page displays the collision and eventual merging of two galaxies. The simulation begins with the two galaxies approaching each other along parabolic orbits. The first passage (at a simulated time of 250 million years) of the galaxies through each other generates strong tidal forces that severely distort the galaxies, throwing out a pair of extended tails. Both galaxies develop bars across their nuclei as a result of this interaction. Another result is orbital decay; instead of traveling away from each other on their original parabolic trajectories, the two galaxies fall back together for a second encounter (at 625 million years). They merge shortly thereafter, leaving a

single object. Note the remarkable similarity of some of the later frames with the observed colliding galaxies shown on the opposite page.

There are many circumstances in which particles are more uniformly distributed than stars in a galaxy. Examples range from hot, ionized gases to the overall distribution of matter throughout the universe. In such cases, the uniformity of distribution sometimes allows researchers to employ an approximation method called a particle-mesh code, which is much faster than summing the effects of individual particles in a strict N-body simulation. The method first calls for a finite difference grid to be established over the region to be simulated. Because the particles are more uniformly distributed, a good approximation can be arrived at by averaging the particles in each finite grid box into a "mass cloud" of uniform density. The program then calculates the force exerted on each particle by each of the mass clouds, rather than calculating the force on each particle by summing the contributions from all the other particles. During each time step, the mass clouds move slightly in response to this computed force, thereby changing the density distri-

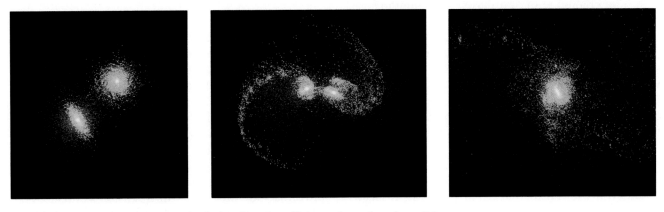

These frames from a supercomputer simulation show the collision and merging of two disk-shaped galaxies. Stars in the disk of each galaxy are colored blue, whereas those in the central bulges are yellow. Red denotes the halos of invisible dark matter that surround each galaxy. Each particle retains its color throughout the simulation. The first image shows the galaxies at 125 million years, and the others follow at intervals of 375 million years.

bution slightly and thus the gravitational or electromagnetic force. In this repetitive fashion, the system is marched forward in time.

To illustrate particle-mesh techniques, we shall consider an attempt to simulate the evolving distribution of matter across vast reaches of the universe. Astronomical observations have recently revealed that clusters of galaxies are found mostly along enormous, roughly spherical surfaces that surround huge voids or along long, thin filaments. This patterning, called large-scale structure, is somewhat like the motif seen in a sink full of soap bubbles. One of the major goals of modern cosmology is to understand the origin and evolution of this cosmic "sudsy" structure.

From observations of the motions of real galaxies, astronomers have deduced that most of the matter in the universe is sufficiently underluminous that it does not show up on photographs. Such material, called dark matter, may in fact compose as much as 90 percent of the matter in the universe. Dark matter possesses mass and exerts a gravitational pull on the luminous matter in galaxies. An understanding of how the dark matter is distributed across space would give us important clues about the large-scale sudsy patterns that clusters of galaxies outline.

The challenge of simulating the changing distribution of the dark matter in the universe over a long period of time was taken up in the mid-1980s by Joan Centrella at Drexel University and her colleague Adrian Melott of the University of Kansas. They used a Cray X-MP at Digital Productions to follow the progress of 64^3 or 262,144 mass clouds over a time long enough for the universe to expand by a factor of 1000.

Because a particle-mesh code is so fast, a supercomputer like the Cray X-MP can handle up to a million mass clouds. The best results are obtained if the number of mesh points is at least as large as the number of mass clouds, so Centrella's group used a computational space consisting of a three-dimensional cube 64 zones on a side. Periodic boundary conditions in all three dimensions ensured that the sides of the cube would not produce artificial effects.

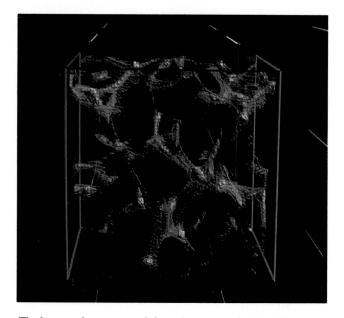

The large-scale structure of the universe was simulated by distributing 64^3 particles throughout a cubic computational domain divided into 64^3 zones. Long-wavelength, low-amplitude density perturbations grew by the action of gravity to produce the filamentary structure seen here. The transparent surfaces show where the local density is twice (blue), four times (green), and eight times (red) the average density of the universe at that moment.

The simulation began when an even distribution of mass clouds throughout the grid was disturbed by a small-amplitude, long-wavelength perturbation in the density. The image on this page, created by Craig Upson of Digital Productions, shows the distribution of matter after a thousand-fold expansion of the universe, at a time comparable to the present. Many of the features seen in the simulation, such as the large voids, are similar to observed maps of the real universe, but there remain quantitative discrepancies between the model and the observations. This research group, as well as many others, has refined its techniques a great deal in the years since this pioneering simulation was done. We shall explore the current state of this research in Chapter 8.

Quantum Systems and the Schrödinger Equation

The modern description of the atomic world originated in the 1920s with the work of physicists like Neils Bohr, Erwin Schrödinger, and Werner Heisenberg, who formulated quantum mechanics. At these extremely small scales a continuous description of matter breaks down, and matter takes on characteristics of both waves and particles.

At the core of quantum theory is an equation, called the Schrödinger equation, whose solution is a "wavefunction" that contains all information necessary to describe the behavior of subatomic matter. For instance, the square of a particle's wavefunction evaluated at a particular point in space equals the probability that the particle is located at that point. The Schrödinger equation is the equation that physicists, chemists, and biologists must solve to achieve an atomic-level understanding of the structure and properties of any material, chemical reaction, or drug.

A wavefunction fully describes how subatomic matter arranges itself in response to the energies of interaction of the quantum particles. The electrons and protons of an atom are charged particles whose interactions are governed by the electromagnetic force. Electrons repel one another on the one hand and are attracted to the protons in the nucleus on the other. These repulsions and attractions together determine the electronic structure or configuration that describes the atom and is given by the wavefunction. The combined electromagnetic interactions of subatomic particles are encoded mathematically in a quantity termed the potential energy, which is inserted into the Schrödinger equation to compute a particle's wavefunction.

The quantum world poses significant computational challenges because of the enormous number of variables that are involved compared to classical computations. To give a specific example, let's consider the storage requirements of 10 charged particles interacting by electromagnetic forces. The classical motions of these 10 particles can be computed in the three dimensions of space by evaluating the forces acting on the particles and then solving Newton's law of motion, as shown in the earlier section on N-body methods. Memory needs to hold six numbers (position and velocity) for each of the 10 particles, or a total of 60 numbers to specify the system at any given time.

In contrast, let us imagine that we want to calculate the quantum electronic structure of a neon atom, which possesses 10 electrons. The potential energy of a particular arrangement of the electrons is obtained by adding up the electrostatic repulsion of every electron by every other, and calculating the attraction of every electron to the positive nucleus. This potential energy is inserted into the Schrödinger equation to become the driving term in the same way that the forces are the driving terms in Newton's equations. But whereas a particle in classical systems is a point with no width, a single particle in quantum mechanics cannot be pinned to a definitive location because of the uncertainty principle. It is described by a wavefunction, which to be uniquely specified requires a value at every point in space. If we discretize the space by imposing a three-dimensional grid having 100 divisions on a side, then a single particle would require 100^3, or one million, numbers to be approximately specified. So, whereas memory holds six numbers for each classical particle, in this case it would hold one million numbers for each quantum particle.

However, solving the Schrödinger equation for an entire system of particles such as the neon atom is not as simple as moving from one to many classical particles. All 10 particles of the neon atom are defined over the same three-dimensional grid, and thus their wavefunctions overlap. Because of the distributed character of the particles' wavefunctions, one must solve for the wavefunction of the entire system to obtain the exact answer to the Schrödinger equation.

That answer depends on the coordinates of all 10 particles, which is a 3×10 dimensional space. That is, when defined on the same grid as above, the wavefunction of the neon atoms requires $100^{30} = 10^{60}$ grid points for its specification. Just storing sufficient numbers is an intractable problem for systems of many particles.

This problematic trait of the Schrödinger equation was pointed out in 1929 by one of the pioneers of quantum mechanics, P. A. M. Dirac: "The underlying physical laws are completely known, and the difficulty is only that the exact application of these laws leads to equations much too complicated to be soluble."

Of course, there are a few exact solutions possible, such as the electron structure of single hydrogen or helium atom or a single small molecule. However, only in the last fifteen years have supercomputers become powerful enough to begin to calculate accurate, but approximate solutions for quantum systems of larger atoms, molecules, or bulk materials. Owing to the computational complexity of the quantum world, by the early 1990s academic users of the National Science Foundation supercomputer centers were consuming far more time in quantum computations than in all of classical science and engineering. This newfound ability to solve the Schrödinger equation numerically touches every field of research that has anything to do with atoms. It has created a powerful and efficient means of theoretical exploration whereby important substances, ranging from exotic alloys to innovative drugs, can be examined in detail at the atomic level to gain fundamental understanding eventually leading to the design of new materials.

Quantum Monte Carlo

In 1946, while the gifted Polish-American mathematician Stanislaw Ulam was convalescing from a near fatal illness, he would play solitaire to occupy his time, and it occurred to him to try to calculate the chances of winning. In other words, what is the probability that all 52 cards would get laid out successfully? After spending a long time fruitlessly trying to calculate the odds, Ulam hit upon the idea of simply playing out a hundred games and seeing how many times he won. Of course, the more games he played, the closer the win/lose ratio was to the value that would be predicted from some complicated analysis. This technique, whereby repeated statistical sampling of a system is used to provide information, is today called the Monte Carlo method, named for the famous gambling resort in Europe.

A variant of this technique, called quantum Monte Carlo, solves the exact Schrödinger equation by using statistical sampling. It is based on methods developed at the Los Alamos National Laboratory during the 1940s by Ulam, Enrico Fermi, and John von Neumann. This mathematical technique can be used to perform simulations of random processes to study problems in quantum physics and chemistry. A widely used version of quantum Monte Carlo, termed Greens function Monte Carlo, was developed in the early 1970s by Mal Kalos, director of the Cornell Theory Center. A simplified variant of this method called diffusion Monte Carlo, developed by David Ceperley and Berni Alder while at the Lawrence Livermore National Laboratory, exploits the fact that the Schrödinger equation can be made to look a lot like a diffusion equation—that is, an equation that describes particles diffusing in a medium by carrying out independent random walks.

Quantum Monte Carlo methods differ from methods for calculating the usual kind of physical diffusion in two interesting ways. The first is that the random walk takes place in many dimensions—30 for the neon atom. The basic procedure is repeated for each electron in a multiparticle configuration such as the neon atom: each of the electron's coordinates in three dimensions is randomly varied at each "snapshot" in time, and for each snapshot the poten-

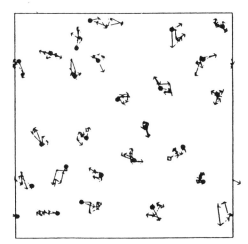 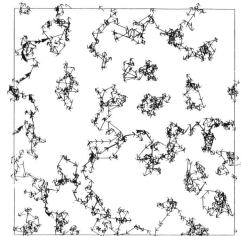

A quantum Monte Carlo simulation reveals the behavior of helium atoms as helium changes from an ordinary fluid to a superfluid. On the left, helium atoms at a temperature of 2.5 K exhibit essentially random motions. On the right, atoms of superfluid helium at 0.8 K follow winding paths indicative of the coherent vortices that develop in a superfluid liquid.

tial energy of the new configuration is computed. By repeating this process enough times, one builds up a statistical ensemble of configurations dominated by many very similar, highly probable configurations. Since these separate "throws of the dice" can be performed independently for each particle, the technique is tailor-made for massively parallel supercomputers.

The second difference is that if a new configuration is found to be too improbable, it is dropped from the ensemble. The probability of a configuration is given by the value of the potential energy of the electrons; the most probable configurations have the least energy. This process of "importance sampling" greatly enhances the efficiency of the method and allows computations to be performed for systems containing up to several hundred quantum particles.

Quantum Monte Carlo has been essential in developing a quantitative theory of liquid helium, which exhibits quantum effects at the macroscopic level. Liquid helium is especially interesting because, at temperatures near absolute zero, it becomes a superfluid that flows without resistance, and indeed can flow uphill. It is known that this strange substance also can have quantum vortices, like miniature tornados, which can spin forever in the liquid. Mal Kalos,

David Ceperley, and their colleagues have used quantum Monte Carlo simulations to understand the behavior of these quantum fluids from first principles.

In one study, Ceperley, now at NCSA and a professor at the University of Illinois at Urbana-Champaign, studied the motions of helium atoms at temperatures slightly higher and slightly lower than the transition temperature for superfluid behavior. The illustration on this page shows the simulated random motions of the individual atoms at two temperatures on either side of the transition temperature. At the higher temperature no coherent motion is seen, but at superfluid temperatures, atomic paths spiral and turn in a way that is characteristic of the vortices in superfluids.

With these techniques, researchers can compute the properties of quantum matter to a high accuracy, taking into account all the quantum mechanical interactions. As we will see in the next chapter, these methods allow scientists to discover properties of quantum materials that are extremely difficult or impossible to observe experimentally, such as the arrangement of atoms in bulk hydrogen subjected to pressures that are millions of times higher than the Earth's atmosphere.

Electronic Structure

With the aid of supercomputers quantum Monte Carlo techniques can generate numerical solutions to the exact Schrödinger equation for configurations of few electrons, but practical applications often involve so many electrons that one must give up finding the complete quantum mechanical solution. For instance, each tungsten atom in a light bulb filament has 74 electrons; thus, to solve for the electronic structure of even a single tungsten atom would require calculating a 222-dimensional space. A cube of atoms with 10 atoms along each dimension, still a very rough approximation of a piece of tungsten metal, would involve several hundred thousand dimensions!

To cope with this unavoidable complexity, scientists have devised clever methods to obtain realistic results without calculating the behavior of every electron and atom. One can imagine computing the wavefunction of a single electron in the average environment created by all the others. The set of one-electron wavefunctions computed in this way gives the electron distribution for one or more atoms. This method would reduce the count of numbers needed to specify the ten-electron neon atom on the million-zone three-dimensional grid from 100^{30} to 10×100^3. When applied to the outer electrons on large atoms, this so-called independent particle approximation can give reliable answers to questions about the properties of the material.

The single-electron approximation does neglect some important physical interactions. No two electrons in a small volume can have identical wavefunctions, and therefore their wavefunctions cannot be totally independent. Although this so-called correlation effect is neglected in computing the one-electron wavefunctions, it can be corrected for after the computation by perturbation theory. Indeed, the accuracy of a simulation often depends on how well this correction is made.

To calculate an electron distribution, you need to know the potential energy that describes the electromagnetic forces produced by some arrangement of atoms. A major difficulty is that, in a multi-atom system, the interactions of the electrons with each other contribute significantly to the potential energy. Since potential energy depends on the electron distribution, you find yourself in a quandary whereby you must already know the answer before you can start to solve the problem.

For many problems in the electronic structure of bulk materials, single atoms, or molecules, one solves this dilemma by computing the approximation for a single electron over and over; each time the solution approaches the most probable distribution a little more closely. First one selects a set of trial "basis functions" to describe a generalized electron distribution. This set comprises simple, exact mathematical functions that are well matched to the geometry of the problem. For bulk materials a popular variant of this approach describes the electron distribution as a sum of unknown coefficients times a set of spherical functions representing the inner core of atoms and a set of plane waves in the region between the atoms representing the outer electrons. If we think of a single-electron approximation as a recipe for an electron distribution, the basis functions are the ingredients and the unknown coefficients give the quantities of the ingredients that will create the particular electron distribution. The unknown coefficients will be solved for to determine the electron distribution for the arrangement of atoms at hand.

To begin, given an arrangement of atoms, you guess a trial electron distribution by giving likely values to the unknown coefficients. You then calculate the potential energy that results from this distribution of electrons and atoms. Using this potential energy in the Schrödinger equation, you compute a more accurate electron distribution. With this improved electron distribution you then recalculate the potential energy, which you use to compute a still better electron distribution. When the $(N + 1)$th iteration gives the same answer as the Nth iteration, you have achieved a self-consistent solution.

This approach has been implemented for solids as the so-called full-potential linearized augmented plane wave method, or FLAPW for short. It is ideal for a supercomputer because it requires solving a huge set of equations over and over. Using the FLAPW method, it is possible to perform highly precise quantum mechanical simulations of bulk solids and their surfaces involving almost any arrangement of atoms. An example from the work of Art Freeman of Northwestern University and Eric Wimmer of Cray Research and their colleagues explores a way to improve the electrical properties of tungsten.

Since the invention of the incandescent light bulb, tungsten has been used as a filament in a wide variety of electrical and electronic devices because of the ease with which it gives up electrons. Tungsten's usefulness was bolstered in the 1920s when the physicists K. H. Kingdon and I. Langmuir discovered that the electron emission rate from the metal is enhanced by the deposition of a thin overlayer of cesium. The presence of cesium atoms lowers tungsten's work function (that is, the energy needed to liberate electrons from the metal), thereby making the cesium-tungsten composite especially successful in such high-tech applications as photoemitters, electron emitters, thermionic emitters, ion propulsion systems, and negative ion sources for magnetic fusion devices.

The reason for cesium's ability to lower the work function took 60 years to be discovered. Using the FLAPW method, Freeman and Wimmer simulated a slab of tungsten five atoms thick covered with a one-atom-thick layer of cesium. The resulting electron distribution is conveniently displayed as a map showing the density of electron charge, seen on page 54. Electrons in the outermost levels of the cesium atoms are pulled toward the tungsten surface,

leading to an increased electronic charge at the cesium-tungsten interface (blue-green) and a depletion of charge on the vacuum side of the overlayer (purple). In addition, electrons deeper inside the cesium atoms are offset in the opposite direction. The net effect of these electric dipoles is to cut the work function nearly in half, from 4.77 electron volts for a clean five-atom slab of tungsten to less than 3 electron volts when the cesium overlayer is added. The FLAPW method allows scientists to quickly determine the most promising combinations of surfaces and overlayers for a variety of applications.

Quantum Chemistry

To understand the electronic structures of individual atoms or molecules or to compute chemical reactions, we must be able to compute all of the interactions of the electrons that occupy the atomic shells. A number of self-consistent, iterative methods, including local density functional methods, are used by thousands of computational chemists for this purpose. Instead of computing the full quantum interactions of all the electrons with each other, as would be necessary for an exact solution, these methods proceed as in the last example by treating each electron as if it were moving in a electrostatic field created by the average of all the other electrons.

These methods are generally referred to as *ab initio* methods, since in contrast to the semi-empirical methods discussed shortly, they do not rely on experimental data to fix some of the terms defining the potential energy. Rather the unknown coefficients for a set of basis functions are solved for, and the electron distribution is recalculated repeatedly until the solution is self-consistent. This process arrives at the system of least energy, while the rates of change near this minimum give important information about atomic vibrations or chemical reaction rates.

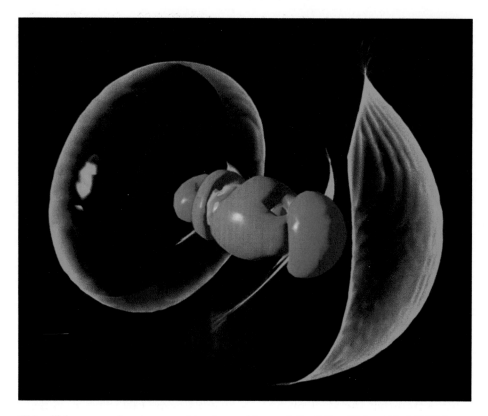

This still from an animation created by the graphics specialist Jeffrey Thingvold of NCSA shows the changes in electron density as the chlorine atom approaches and bonds with the chromium atom in a computation by Aileen Alvarado-Swaisgood. The red surface encloses volumes in which the electron density increases, the blue surface encloses regions in which the electron surface decreases, and the green surface encloses regions of no change.

The most widely used *ab initio* method dates from the 1930s and bears the names of its originators, Hartree and Fock. These techniques are at the core of widely used, publicly available codes such as the "Gaussian '92" that have evolved from codes developed in the 1960s by John Pople of Carnegie Mellon University and his colleagues. The one-electron wavefunction, called a molecular or atomic orbital by chemists, is used to compute terms in the potential energy describing electron-electron interactions. The number of these interaction terms grows as roughly the fourth power of the number of one-electron wavefunctions.

When these methods were first developed, the interaction terms were stored on rotating disk and then read back into the computer for every iteration. But in the last decade, as super-computer speed increased faster than disk performance, Jan Amlof of the University of Minnesota and his coworkers pursued a direct self-consistent method that took advantage of the increased speed to simply recompute the terms each time they were needed. Now as multi-gigaword memories become possible, the execution time may be speeded up yet again simply by storing the terms in memory. Once again, we see the interplay between machine architecture and computational algorithms. For any of these approaches, one can improve the accuracy of these methods by adding more refinements to the physics, such as corrections for correlation effects.

The work of Aileen Alvarado-Swaisgood of Amoco Oil Company illustrates what these methods can accomplish. Using *ab initio* methods on a Cray-2 supercomputer at NCSA, she and her colleagues collaborated with NCSA chemist Harell Sellers to study the catalytic properties of a chromium chloride molecular ion as it bonded with a chlorine atom. The strategy was to visualize the shifts in the electron distributions as the distance narrowed between the approaching chlorine atom and chromium ion.

The image on the opposite page shows that as one might expect, the bonding has little effect on electrons at large distances from the core of the atoms. The bonding itself is accomplished by a transfer of electron density from regions near each atomic core to the region between the cores. These results led to new insights about how such molecular ions can act as catalysts for organic molecules such as alkanes.

Molecular Dynamics

As we move to molecules larger than a few hundred atoms, such as the large biological molecules we will study in Chapter 5, the computing power required to compute their electron structure by *ab initio* methods becomes prohibitive. Fortunately, one does not have to compute the electron distribution in detail to study many of the important properties of these molecules. For instance, a scientist may wish to study the motions of atoms as parts of a molecule twist or wave or vibrate. Many of the motions of the atoms that make up molecules can be computed using a hybrid of classical and quantum methods.

These methods, termed molecular dynamics, describe a complex molecular system as a set of point masses (the atoms) moving in a force field caused by many atom-atom interactions. In a multi-atom system, the potential energy that mathematically encodes these forces would include mathematical terms describing various types of interactions such as covalent bond stretching, bond angle bending, and torsional angle distortions, which correspond to the various ways in which molecules vibrate, bend, and twist. The values of these terms are determined either by adjusting parameters to fit experiments or by performing quantum chemical calculations. Once the force fields have been defined, one can then solve an N-body problem, subject to Newton's laws of motion, with the methods described on pages 73 to 76.

Assuming one has computed the trajectories of the atoms in the molecule, how does one visualize and analyze the results? One can form statistical averages of the bulk motions, but some of the details of the motions may be lost. In many cases, interactive graphics workstations display the oscillating and wiggling molecules so that the researcher can try to grasp general trends. However, remembering the history of all the moving side branches is a daunting task.

Recently, David Herron of Eli Lilly and Company and graphics specialists William Sherman, Jeffrey Yost, and Jeffrey Thingvold at NCSA devised some new techniques as part of Lilly's efforts to develop new drugs that could ease the suffering of asthma patients. A drug with the appropriate molecular shape would inhibit the cellular binding of a family of molecules called leukotrienes, or LTE for short, that are implicated in asthmatic conditions.

Using the commercial molecular dynamics package Amber running on a Cray-2, Herron compared the movements of atoms within three variants (termed LTD_4, LTC_4, and LTE_4) of LTE. These molecules, shown at the top of the next page, have a common core structure, but differ in the shape of their tails to the left and

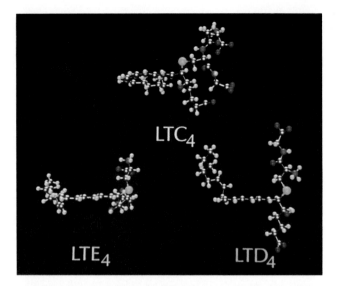

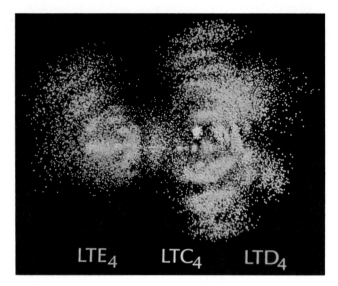

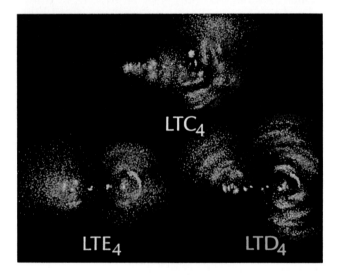

Top left: *The three variants of LTE in their static equilibrium configurations. Key atoms are color coded.* Bottom left: *The spatial extent of the vibrations around equilibrium for each of the three molecules.* Above right: *A superposition of the three molecules' vibrational excursions. LTE_4 is pink, LTC_4 is yellow, and LTD_4 is blue.*

the right. For the simulation, each molecule was placed in a bath of water molecules to replicate its biological environment, and its atomic motions were computed. By shrinking the visual size of the atoms, and summing up the movements over many time steps for each molecule, Herron's team created images displaying the extent of the molecular motions, shown above.

Finally, these three figures can be superimposed, with color now representing the label of the molecule. Now that Lilly's scientists know the "generic" shape of the family of moving molecules, they have a much better chance of creating an inhibitor drug. This information is crucial in the process of the rational drug design, a procedure we will return to in Chapter 5.

A variant of the molecular dynamics approach allows one to compute the electronic structure of a molecule or a material while atoms are in motion, rather than determining just one static structure as local density functional methods do. These techniques, such as those published in 1985 by Roberto Car of the International School of Advanced Studies in Trieste, Italy, and Michele Parrinello of IBM Research Laboratory in Zürich, Switzerland, are beginning to make possible major breakthroughs in the calculation of the structure of materials.

For instance, in 1990, Car, Parrinello, and Parrinello's IBM colleague Giulia Galli, with Richard Martin of the University of Illinois at Urbana-Champaign, used these methods on NCSA's Cray Research supercomputers to study the melting of diamond at high pressures, an effect first directly observed experimentally only in 1984. They found that, contrary to what had been expected by analogy with silicon or germanium, the melting temperature increases with pressure.

More recently, in early 1992, two groups, one from England and one from MIT, AT&T, and Thinking Machines Corporation, used Car-Parrinello techniques on massively parallel supercomputers to compute the lowest energy of a specific surface configuration of silicon, one of the most complex and widely studied surfaces of a solid, because the chemistry of silicon surfaces is the basis for the semiconductor industry. This tour de force of a calculation followed the motions of nearly a thousand atoms to accurately describe the atomic bonds and energies of interaction. In the future, it is thought that many of the semi-empirical force fields used in computing the molecular dynamics of biologically important molecules can be replaced by force fields calculated from first principles using supercomputing techniques such as Car-Parrinello.

In the chapters to come, we will follow the stories of researchers as they use the approximation methods of supercomputing to attack an enormously wide range of topics. We have ordered the chapters in ascending scale, from the smallest of subatomic particles to the vast universe itself. Each story is chosen to be representative of the field, but we could just as well have chosen any of the many other thousands of stories from the last few years of supercomputers successfully applied to the tasks of scientific discovery and engineering design.

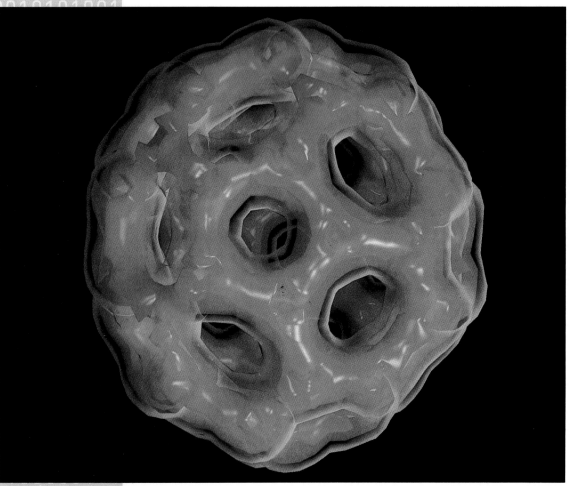

The Quantum World

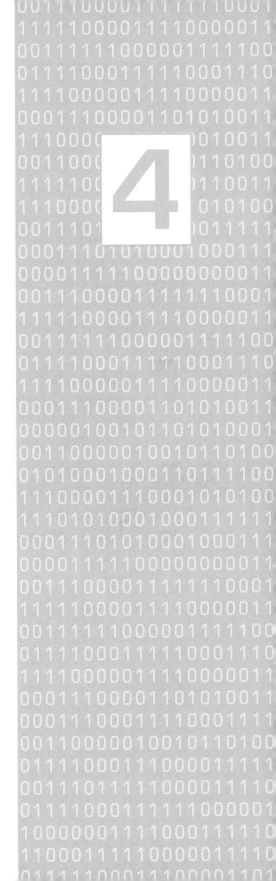

4

A supercomputer programmed with the laws of physics can simulate the atomic and subatomic phenomena that give rise to the material universe. Its portal to the subatomic world is the Schrödinger equation, the fundamental law governing quantum mechanics. Solutions to this equation are wavefunctions for quantum particles, in keeping with the duality of wave and particle that is a fundamental, inescapable feature of quantum mechanics. In spite of the famous uncertainty associated with quantum particles, each one possesses a definite quantum state specified by the wavefunction given in the solution to the Schrödinger equation.

The colors yellow, green, and blue denote regions of successively greater electron density in the 60-carbon-atom molecule buckminsterfullerene.

When we move from consideration of a single particle to bulk matter containing many particles, some surprising differences arise between different classes of quantum particles. Besides possessing the features that particles have in the classical world, such as mass and energy, all quantum particles also possess what physicists call "spin," one of several arcane properties that together define a particle's quantum state. One of the major divisions of the quantum particle world is determined by the value of the spin. Quantum particles in one great family have an integer spin, while the particles in the other family possess half-integer spin. Those having integer spin are the "bosons" such as the familiar photon, the elementary particle of light. Bosons can condense in great numbers into the same quantum state, thus making possible such technological wonders as the laser.

Those particles whose spin is a half-integer, called fermions, include the constituents of atoms such as electrons, protons, and neutrons. These particles behave very differently from bosons because they obey a law of physics called the Pauli exclusion principle, which was formulated by Austrian physicist Wolfgang Pauli in 1925. The principle tells us that no two fermions can occupy the same quantum state at the same time. Indeed, the fermions "exclude" each other, providing an effective "force" that "stacks up" the fermions when they find themselves close to each other.

This seemingly esoteric principle is the fundamental reason why the world of matter exists in the forms we know it. It is critical to the manner in which quarks combine to form protons and neutrons, and it determines how different numbers of protons and neutrons create the various nuclear isotopes. It requires electrons in solid materials to behave in different manners to create conductors, insulators, and semiconductors; it makes electrons distribute themselves in distinct shells in atoms, creating elements of differing chemical properties; and it provides for the rich variety of structure in the molecular world of chemistry.

The Elementary Particles

In their quest to discover the basic building blocks of matter, physicists search for objects they call elementary particles. Such particles have no internal structure and behave like infinitesimal points of mass and electric charge. The electron is one such particle, but the protons and neutrons that make up the atomic nucleus are not. Experiments using particle accelerators reveal that protons and neutrons have a finite size (their diameters are roughly 1 fermi = one-millionth of a nanometer, where a nanometer is one-billionth of a meter!) and contain three pointlike particles called quarks. Quarks are believed to be truly elementary particles in the same sense that electrons are. Like electrons, the quarks have half-integer spin and therefore are fermions.

In the 1970s, a new law of nature called quantum chromodynamics, or QCD for short, emerged that explains how quarks interact by exchanging another class of particles whimsically called gluons. Gluons are the fundamental carriers of the strong force that binds quarks together. The theory of quantum chromodynamics is modeled on the theory of quantum electrodynamics, which describes how photons (the carrier of the electromagnetic interaction) interact with particles having electric charge. These mathematical formulations of laws of nature are called gauge field theories, and the carriers of the forces (like the photon and the gluon) are bosons called gauge particles.

All elementary particles other than quarks that have electric charge, the source of the electromagnetic force, possess integer multiples of the electron's charge, whereas the electric charge of quarks comes in units of one-third of the electron's charge. In addition to electric charge, quarks carry another "charge," termed color, which is the source of the strong force, hence the term "chromodynamics." Color, which is a nonsense word having nothing to do with real color, can be thought of as coming in three shades: red, green, and blue. Finally, there are six "flavors" of quarks: up, down, strange, beauty, bottom, and top. The mass of the quark is much greater as one goes through the list from "up" to "top."

Unlike photons, which carry no electric charge themselves, gluons possess color. This small difference makes the force between quarks behave very differently from the well-known electric force, which is very strong when two electric charges are close, but which dies off as the distance between the electrically charged particles increases. In contrast, quarks and gluons interact weakly at distances that are small compared to the diameter of a proton, but interact strongly at larger distances. In other words, quarks are free to rattle around inside protons and neutrons, but if one of the quarks tries to wander off, the pull on it by the remaining two is so strong that the proton or neutron remains intact. This phenomenon is termed quark confinement.

The concept of quarks originated with the work of the physicists Murray Gell-Mann and George Zweig in the early 1960s. At that time, physicists were quite puzzled by the profusion of exotic particles created during the collision of protons and other subatomic particles in particle accelerators. Gell-Mann and Zweig made sense of this plethora by hypothesizing that all the known heavy particles affected by the strong interaction, collectively called hadrons, could be understood as a combination of either two or three elementary particles that Gell-Mann called quarks, from a nonsense word made up by James Joyce.

The proton and neutron, and their ephemeral cousins that exist for a tiny fraction of a second during accelerator experiments, all form the "baryon" subfamily of the hadrons. Each baryon consists of three quarks, and in the proton and neutron each of these quarks has the flavor up or down. All the heavier baryons that are stable with respect to the strong interactions have at least one quark of flavor other than up or down. For each type of baryon particle, there is a corresponding antiparticle of equal mass but

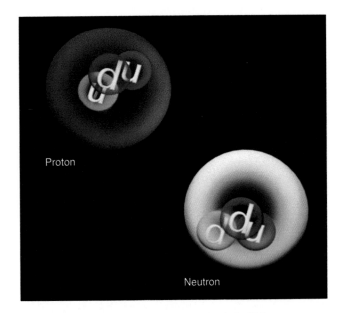

Proton

Neutron

Three quarks of different color create a "colorless" baryon. Protons and neutrons are each made of three quarks of different color whose flavors are up and down. The fractional electric charges of the quarks (up = $\frac{2}{3}e$, down = $-\frac{1}{3}e$) add up to the single charge (e) on the proton (up, up, down) and the zero charge on the neutron (up, down, down). This computer graphics visualization was created by Matthew Arrott of NCSA for use on the Japanese public broadcasting system NHK.

opposite electric charge, such as the antiproton or antineutron. The baryon antiparticles are made of three antiquarks.

The second subfamily of the hadrons, known as mesons, are composed of one quark and one antiquark. For instance, the lightest meson is called the pion, postulated by the Japanese theorist Hidekei Yukawa in 1935 as the carrier of the nuclear force binding neutrons and protons inside the nucleus and discovered experimentally in 1947. The meson-mediated nuclear force is now seen to arise entirely as a side effect of the strong force interactions between the quarks and gluons inside the hadrons.

Using quantum chromodynamics, physicists should, in principle, be able to compute all the properties of hadrons from the properties of their internal quark constituents such as mass charge and flavor and from their gluonic interactions. However, because the gluon force is so strong, physicists cannot calculate exactly or with perturbation theory the properties of quarks bound by gluons inside hadrons. In 1974, Kenneth Wilson, later to become a Nobel laureate and the founder of the NSF supercomputer center at Cornell University, showed that a discrete spacetime lattice formulation, which preserves the underlying symmetries of the exact theory, naturally emerged for studying QCD.

In this formulation of QCD, called lattice gauge theory, the wavefunctions of the quarks reside on the vertices of the grid, and the gluon wavefunctions are defined on the links between the vertices. The computation requires an elegant formulation of quantum mechanics developed by Richard Feynman. The so-called Feynman path integral formulation computes probabilities by summation over all possible classical paths the system can take. In QCD, specifying a "path" means specifying the values of the quark and gluon wavefunctions at each point in spacetime. Using quantum Monte Carlo or molecular dynamics techniques similar to those described in Chapter 3, one builds up an ensemble of the configurations of the quarks and gluons in a particle, from which can be computed accurate physical averages of particle characteristics such as radius and mass spectrum as well as bulk characteristics such as density and pressure.

If the complexity of the calculations required to specify the wavefunction of an electron seemed baffling in Chapter 3, then the complexity of QCD calculations is even more extreme. Although the wavefunction of an electron has to be defined at each point on the entire spacetime grid, there need be only one or a few numbers at a point. In QCD, establishing the value at each point on the grid requires hundreds of numbers. Thus computation times can be orders of magnitude longer for a grid of comparable size. Because of this complexity, in the seven years since the NSF supercomputer centers opened, simulations of QCD have used well over one hundred thousand hours of supercomputer time!

The simplest question a simulation can answer is whether solutions to QCD obtained by using lattice gauge theory can produce the ordinary hadrons such as protons or pions. The wavefunction that emerges from such computations should give us the distribution of quarks in a hadron. For instance, Thomas DeGrand and M. Hecht at the University of Colorado have recently done such a computation on the massively parallel Thinking Machines CM-2 at the Pittsburgh Supercomputing Center. Using Monte Carlo methods on a $16 \times 16 \times 16 \times 32$ spacetime grid, they computed thousands of "snapshots" that show different configurations of quarks inside hadrons, based on statistical probability. By averaging these, one finally obtains the spatial distribution of the wavefunction of the quarks.

The proton wavefunction is constructed from two up quarks and one down quark. One way to quantitatively visualize the wavefunction is to hold the up quarks fixed and plot the wavefunction of the down quark. In the visualization on the facing page, one can see the extended character of a proton, even though only one of its three quarks is visualized. In addition, the work of DeGrand and Hecht has yielded the wavefunctions for the quark and antiquark in the pion and for the quarks in the excited states of both particles.

DeGrand and other workers in this field face a number of technical difficulties in working out the solutions. First of all, real up and down quarks are quite light, typically having masses only a few tenths of a percent of the proton's mass. The lighter the quarks, the longer the computation time, so the best DeGrand and others can do is use masses that are several times too heavy. A second problem is that it is very much more time consuming to include in the computation all the ramifications of the ex-

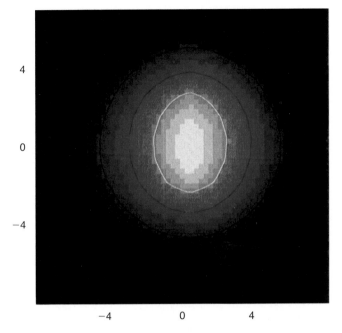

The wavefunction of the down quark in the proton is plotted as the two up quarks are held fixed on the horizontal axis at positions (0, −2) and (0, 2). The colors indicate the value of the wavefunction: white is the highest, the yellow shades are intermediate, and red shades are lowest. The square of the wavefunction represents the probability of finding the quark at that position.

clusion principle obeyed by the fermionic quarks. Finally, the lattice spacing is still too coarse to permit accurate comparison with experiments.

Scientists would like to work out all the properties of individual hadrons: their mass spectra, radii, magnetic moments, lifetimes, interactions, and so on. Although possible in principle, these computations will require supercomputers far beyond today's power. Even teraflop computers may be too slow. Fortunately, the exponential increase in supercomputer speed should continue well into the next century. By the time the next great particle accelerator is built, lattice gauge theory will most likely be a routine tool used in conjunction with the experimental detection of particle collisions.

Hadronic Matter

Under certain extreme conditions of heat and density, as at the core of a neutron star or in the early universe, nuclei are squeezed together to create a strange form of matter consisting only of hadrons. Hadronic matter can come in a remarkable variety of physical states, or phases, that are analogous to, though different from, the states of solid, gas, and liquid familiar from more ordinary kinds of matter. The phase of hadronic matter changes according to the temperature and the net baryonic density, which is computed by subtracting the number of anti-baryon particles from the number of baryon particles in a given chunk of hadronic matter. Exploring the characteristics of bulk hadronic matter is another "grand challenge" of lattice gauge theory.

One state of hadronic matter is exhibited by the roughly spherical concentration of protons and neutrons in the normal nuclei of atoms. As we will see in Chapter 8, a much larger body of hadronic matter in this state is created when the core of a massive star collapses to form a neutron star. The collapsed core is, in effect, a gigantic nucleus with a mass greater than the mass of our Sun. Both in atomic nuclei and in neutron stars, the quarks remain safely inside the neutrons or protons. However, their confinement may end if conditions become extreme enough. Several nuclear physics laboratories have built relativistic heavy ion colliders that smash large nuclei (up to uranium!) against each other, creating for a moment an extremely dense and hot nuclear fireball. Inside this microscopic fireball, the nuclei are so squeezed together that the quarks and gluons become liberated from their nuclear bags and form a "quark soup."

When the universe was less than one microsecond old, all matter was so compacted that the whole universe was full of quark soup. Unlike conditions in the core of a neutron star, the temperatures were unimaginably high, well over

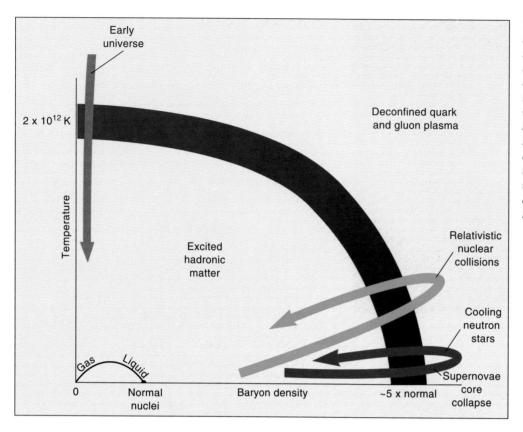

This phase diagram summarizes the possible states of hadronic matter and the corresponding conditions of density and temperature. Ordinary nuclei with no antiparticles present lie partway up the horizontal axis of the diagram at near zero temperature. The circular arc indicates the liberation of quarks and gluons at very high temperature or density.

a trillion degrees Kelvin. As the universe expanded, the density lowered to the point that the quarks could segregate themselves to form baryons and antibaryons. Since these particles were all crammed together, all but one in ten billion of the baryons collided with antibaryons, annihilating both particles and forming photons. This annihilation left the early universe with a net baryonic density of almost zero as it crossed the critical temperature at which matter condensed into baryons.

The detailed nature of the transition from a quark-gluon plasma (the formal name for "quark soup") to confined quarks is the subject of a great deal of study by researchers in lattice gauge theory. Massive calculations have been carried out by large teams of researchers on both vector multiprocessor supercomputers and massively parallel machines such as the Intel

Delta and the Thinking Machines CM-2. For example, the "High-Temperature Monte Carlo Collaboration," formed of ten scientists from all over the United States, used large amounts of time on the CM-2 at the Pittsburgh Supercomputing Center, sustaining speeds in excess of 2 gigaflops. The members of this collaboration include S. Gottlieb and A. Krasnitz from Indiana University, U. M. Heller and A. D. Kennedy of Florida State University, W. Liu of Thinking Machines Corporation, J. B. Kogut and R. L. Renken of the University of Illinois at Urbana-Champaign, D. K. Sinclair and K. C. Wang of Argonne National Laboratory, R. L. Sugar of the University of California at Santa Barbara, and D. Toussaint of the University of Arizona. Their studies show that the shift to a quark-gluon plasma under conditions of increasing temperature or density may not be a true phase transition, but a gradual liberation of confined quarks. The group will be among the pioneering users of the sequence of machines that will move us toward and beyond teraflop machines, as will be necessary to achieve a definitive description of the transition.

We are gaining experimental insights into the states of hadronic matter from the new generation of relativistic heavy ion colliders. These colliders take advantage of the fact that positively charged particles (protons) exist in atomic nuclei. Thus when nuclei are placed in an electric field, they are accelerated to speeds extremely close to the speed of light. Because an object gains energy by increasing its velocity, these high-speed nuclei carry a tremendous amount of the energy of motion, called kinetic energy. Indeed, the kinetic energy of an individual nucleus is much greater than its rest mass.

When the nuclei collide straight on, they are brought to an abrupt stop, and all that kinetic energy is converted into the rest mass of newly created particles—mostly excited mesons—that fly away from the collision fireball at high velocity. Incredibly, the fireball lasts only a few ten billion trillionths of a second! Although nuclear physicists have created detection instruments to measure the energy and angle of motion of this debris, there is no way for the experimentalists to measure directly what goes on at the fireball's center. For this reason, they turn to supercomputer simulation of the laws of relativistic nuclear physics to obtain a clearer picture of what happens in this tiny fraction of a second to produce the spray of particles.

To simulate such a collision using supercomputers, one must take great pains to discretize the equations of motion accurately so that they faithfully represent the effects of Einstein's theory of special relativity, especially the strange distortions of space and time that occur at relativistic speeds. Recently, James R. Wilson and Thomas L. McAbee of Lawrence Livermore National Laboratory have developed such a code that treats the hadronic matter in approximate form as though it were a fluid. The equation that specifies the state of hadronic matter is based on work of Vijay Pandharipande, a profes-

The equations of special relativity allow one to compute and visualize the strange distortions that would appear in a regular cubic lattice (left) as it approaches the viewer at 99 percent of the speed of light (right).

sor at the University of Illinois at Champaign-Urbana, and his coworkers. The development of this equation of state has itself required large amounts of supercomputer time. The meson production from the collision is computed using classical Monte Carlo techniques. This version of the Monte Carlo methods treats the motions of particles "semiclassically": the motions are calculated as the result of classical relativistic forces, but the values of the forces are adjusted to account for quantum effects.

McAbee and Wilson's code reproduces well the number and energies of the pions observed

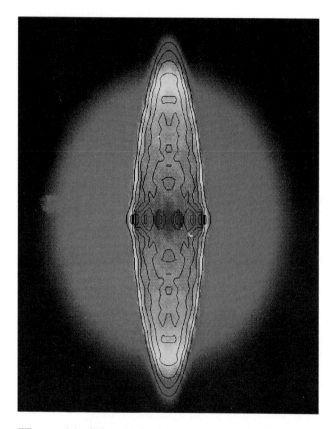

When nuclei collide, a vertical shock wave is created at the plane of impact. The wave moves rapidly outward through the unshocked and still incoming nuclear matter shown in blue. The yellow and red regions inside the shocked pancake have been shock heated to five times nuclear density.

to be flying out of the nuclear fireball region in the collision experiments. By repeatedly computing the equation of state using different values for a sensitive parameter called the Landau-Migdal parameter and then comparing the code's predictions with experiment, McAbee and Wilson were able to refine the likely value of that parameter. Wilson and his colleagues then used the refined equation of state in another computer code to compute the explosion of a supernova and the formation of a neutron star. The explosion produced by the refined equation produces a more energetic lift off of the star's outer layers more reliably than the one produced by the older and less accurate equation of state.

We now move from the fascinating, but rather abstract, world of elementary particles and nuclear physics to the realm of materials that we can see and touch. There, we find supercomputing at the heart of efforts to understand and design new materials atom by atom.

Conductors, Insulators, and Superconductors

Modern technology has given us the ability to develop highly specialized materials that have been significantly modified from their natural state—plastics, metallic alloys, chemicals, and fibers of all sorts. At the microscopic level, the desired characteristics of these materials depend on the arrangement of the atomic nuclei and the distribution of their electrons. By rearranging these basic building blocks, we can in principle create materials with a wide range of desirable properties, such as a preferred electrical conductivity, magnetic attraction, melting point, hardness, or chemical reactivity.

Since the physical and chemical properties of a material are governed by the behavior of its electrons and the spatial distribution of its atoms, the ability to design exotic compounds with a supercomputer is ultimately based on the precision and completeness with which researchers can solve the Schrödinger equation. New materials with desirable characteristics are constantly being discovered, yet because we lack an atomic level understanding of their behavior we cannot modify them except by trial and error. If by computation we could develop a detailed model of the atomic properties of the substance, we could use the supercomputer to explore the results of making alterations to the material's structure.

When a large number of atoms condense into a small volume, as in a liquid or solid, then quantum effects become important. The macroscopic properties of matter result from an interplay between the electrical interactions of the electrons and the ions on the one hand and the Pauli exclusion principle on the other. Matter in its rest state tends to remain in the configuration of lowest energy, but the Pauli principle compels the electrons of matter in this "ground state" to "stack" in levels of differing energy. Because the Pauli principle forbids two electrons from occupying the same quantum state, they cannot all be in the lowest energy state.

When the electrons are in an isolated atom, then they form the familiar "atomic shells," or electron energy levels, that determine the structure of the periodic table of the elements. When energy is added to the atom, an electron is excited and it jumps to a vacant state of higher energy. But the outer electrons of the tightly packed atoms in solids have strong interactions with neighboring atoms, causing the discrete electron energy levels of the atoms to be smeared out into what are called energy bands. Just as there is an energy gap between an atomic energy level and its first excited state, so there can exist energy "band gaps" in solids. The detailed structure of these energy bands and gaps determines whether a solid acts as a conductor, semiconductor, or insulator. These three types of materials are interesting in their own right, but from this book's point of view they are especially interesting: they are the principal components of any digital computer.

All atoms consist of nuclei surrounded by a cloudlike distribution of electrons. Those electrons close to a nucleus are tightly bound and thus do not participate in the conduction of electricity. But the outermost or "valence" electrons of an atom are loosely shared by all the nuclei, and in conducting materials they are free to flow in an electric current. Because the nuclei and their tightly bound core electrons together have a net positive charge, they are often referred to as ions. When studying substances that conduct electricity, scientists must treat the free electrons that carry the current according to the rules of quantum mechanics, which explain that low-energy bands are occupied with electrons, but high-energy states are empty. The electrons in the highest-occupied energy state define the so-called Fermi surface of the material. The conduction of electricity involves electrons with energies at the Fermi surface, often called the conduction energy band, moving through the atomic lattice.

Atomic nuclei are tens of thousands of times heavier than single electrons and thus much less mobile. Although in calculations of the structure of the material they are treated as if they are at rest in the lattice structure of the material, they can vibrate at low frequency around their equilibrium positions. According to the laws of quantum mechanics, an atom can vibrate only in prescribed directions at prescribed energies, and these quantized vibrational modes are called phonons. Since the motion of one ion will affect the neighboring ions, phonon excitations commonly propagate throughout the lattice of a solid. As we shall see, couplings between the electrons and the phonons can lead to exotic behavior in some substances.

For over a century, metals such as copper have been used to conduct electricity in everything from toasters to supercomputers. The flowing current loses energy to heat, however, because of the resistance of the conducting material. In a metal like copper, the mechanism of resistivity has been understood for many years: as the electrons try to move through the metal, they scatter off the phonons. The hotter the metal is, the more vibrations there are, resulting in higher resistivity or, equivalently, lower conductivity. Ordinary metals exhibit a resistivity that is directly proportional to temperature. Insulators, in contrast, do not allow for the free flow of electrons; indeed, their resistance increases as the temperature is lowered. In between these two extremes are the semiconductors, such as silicon or gallium arsenide.

A variety of unusual materials called superconductors offer the possibility of electrical conductance without resistance. The phenomenon of superconductivity was first noticed in 1911 by the Dutch physicist Heike Kammerlingh Omnes, who discovered that as mercury was cooled to near absolute zero, its resistance abruptly dropped to zero. The temperature at which the material loses its resistivity and becomes a superconductor is called the critical temperature (T_c). Below this temperature, a current set in motion will flow forever. Superconductivity therefore holds out the promise of transmission lines that can carry electric power over great distances without any dissipative losses or extremely efficient motors with coils of superconducting wire able to store huge currents indefinitely.

Nearly half a century elapsed between the discovery of superconductivity and its theoretical explanation for substances with a critical temperature near absolute zero, the only type of superconductor known until recently. These low-T_c substances such as mercury, lead, and niobium become superconductors because of the collective action of electrons described in the "BCS theory" published in 1957 by John Bardeen, Leon N. Cooper, and J. Robert Schrieffer. According to this theory, free-roaming electrons near the Fermi surface disturb the phonons produced by the lattice of atoms, creating a force that overcomes the mutual repulsion of the electrons caused by their negative charges. As a result of the phonon mediation, electrons form a macroscopic quantum state made up of what are termed Cooper pairs. Unlike single electrons,

pairs in this state cannot scatter off the phonons, because the energy cost of the resulting breakup would be too high. Hence, resistance is eliminated and superconductivity results.

Can Hydrogen Become a Metal?

Typically, one thinks of conductors or superconductors as metals, such as silver, copper, or niobium. However, given that the electrical properties arise from the electron and lattice structure, an interesting question is whether substances formed of much simpler atoms can exhibit such properties. Amazingly, supercomputer simulations strongly suggest that material composed of the very simplest atom, hydrogen, with its one electron and one proton, can perform both as a conductor and as a superconductor—if it is compressed to high enough pressures. At normal pressure and room temperature, hydrogen is a well-known gas formed of molecules containing two hydrogen atoms. However, as the phase diagram on the facing page shows, at varied temperatures and pressures hydrogen can assume a wide variety of material forms. If we heat hydrogen at normal atmospheric pressure, we eventually reach a point, at 5000 K degrees, where the thermal energy becomes so great that it breaks the molecular bonds and leaves a gas composed of single atoms only. At temperatures above 10,000 K, the electrons are stripped off the proton nucleus, and a plasma of free electrons and protons results.

On the other hand, if we lower the temperature at atmospheric pressure, the molecular gas goes through a triple point at 15 K degrees above absolute zero, where the gas, liquid, and solid states coexist. At higher pressures, a molecular liquid form of hydrogen exists; this liquid is the likely core component of the large planets such as Saturn and Jupiter. Below the triple point temperature, hydrogen exists as a molecular solid. This solid is an electrical insulator over a wide range of pressures at low temperature.

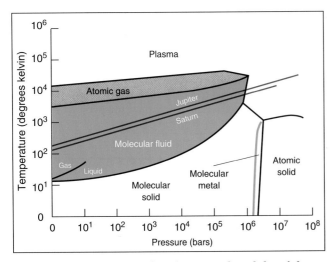

This phase diagram summarizes the current knowledge of the many different material phases that hydrogen can assume. Note the path of pressure and temperature predicted for a cross section through the large planets Jupiter and Saturn. The transition from a molecular insulating solid to a molecular metal to an atomic solid occurs in the horizontal axis at just over one million bars pressure. (A bar is roughly equivalent to an atmosphere.)

However, more than fifty years ago, it was predicted that hydrogen would become a metal at high enough pressures (near a million times atmospheric pressure).

Supercomputer simulations over the past five years have studied hydrogen's transition from insulator to metal in detail. Because of the small number of quantum particles involved, quantum Monte Carlo methods can solve the exact Schrödinger equation. Using the Cray Research supercomputers at Lawrence Livermore National Laboratory, David Ceperley and Berni Alder were able in 1987 to map the ground-state configuration of solid hydrogen at high pressures.

The advantage of their approach is that it can yield in principle exact quantum effects of both the protons and electrons by treating the hydrogen as a two-component system of charged particles of unequal mass (the proton is 1836 times more massive than the electron). The dis-

advantage of the Monte Carlo approach is that only a finite number of atoms, in the hundreds, can be computed in a reasonable time. As a consequence, it is difficult to compute large-scale properties, such as the crystal structure of least energy.

Another approach to this problem, used by Art Freeman and others, is the local density functional method. Working with colleagues in 1989, Marvin Cohen, a professor at the University of California at Berkeley, explored the transition with the aid of Cray Research supercomputers at Livermore and at the Minnesota and Pittsburgh supercomputer centers. Cohen's team used the local density functional approach with quantum corrections adopted from the Monte Carlo work of Ceperley and Alder.

They found that, as the solid molecular hydrogen is crushed under extreme pressure, the energy band gap that makes molecular hydrogen an insulator is reduced to zero, and, for pressures in the millions of atmospheres, a molecular metal is formed. As the pressure rises above 10 million atmospheres, the molecules in the solid break up, turning the hydrogen into an atomic solid. Their computations of this transition showed that the molecular metal has a hexagonal, close-packed lattice structure, which shifts to a more compact rhombohedral structure in the atomic metal phase. Steven Louie and colleagues at Berkeley have since used a more refined method on the hexagonal lattice and report that the phase transition to a molecular metal should occur at 1.5 million atmospheres.

Most interesting, Cohen and colleagues were able to compute the phonon-mediated coupling of electrons into Cooper pairs predicted by the standard BCS theory of superconductivity for metallic hydrogen. They found that at these enormous pressures, hydrogen becomes a superconductor at the surprisingly high temperature of 230 K—nearly twice as high as the critical temperature of the "high-temperature superconductors" made of copper oxide that we will discuss below!

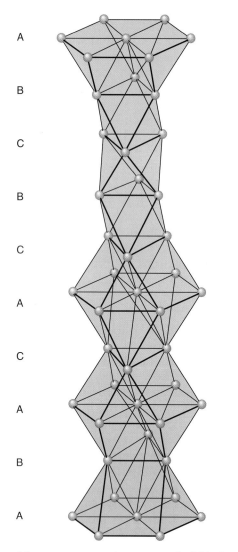

A

B

C

B

C

A

C

A

A

B

A

The computed lowest energy atomic structure of solid hydrogen consists of a rhombohedral structure called "9R," according to Marvin Cohen and Troy Barbee III. Each site is occupied by a single hydrogen atom.

Remarkably, laboratory experiments are now capable of producing these astonishingly high pressures by compressing a minute sample of hydrogen between two gem-quality diamonds. Although the initial results are still controversial, there is strong evidence that a phase transition does occur at 1.5 million atmospheres of pres-

sure. It has not yet been established experimentally whether this is the metal-insulator transition predicted by the supercomputer simulations or whether the new phase behaves as a superconductor. However, we can hope that the intense work, both computational and experimental, now underway will resolve the debate soon.

The 1-2-3 of High-Temperature Superconductivity

The standard superconductors known for many years have led to many useful devices, but they are hampered by the expense and difficulty of using liquid helium (4.2 K) to cool a substance below its critical temperature near absolute zero. In 1986, physicists K. Alex Müller and J. Georg Bednorz at the IBM Research Laboratory in Zürich, Switzerland, discovered that a ceramic compound of the elements lanthanum, barium, copper, and oxygen ($La_{2-x}Ba_xCuO_4$) becomes superconducting at 35 K ($-238°C = -396°F$), half again as high as the previous record-holding critical temperature. Their discovery inspired a worldwide "gold rush" of research to find more "high-T_c" superconductors. Soon the search had resulted in the synthesis of substances that are superconducting at considerably higher temperatures.

The Nobel prize winning work of Bednorz and Müller pointed the way toward substances that are superconducting above the boiling point of liquid nitrogen (77 K), which is much cheaper and easier to work with than liquid helium. By early 1987, Paul Chu and colleagues at the University of Houston and the University of Alabama had indeed discovered an yttrium-barium-copper-oxygen compound ($YBa_2Cu_3O_7$) having a critical temperature of 93 K. Because of its chemical subscripts, this compound soon became known as the 1-2-3 superconductor. A year later, a group at the University of Arkansas unveiled a thallium-barium-calcium-copper-oxygen compound ($Tl_2Ba_2Ca_2Cu_3O_{10}$) that

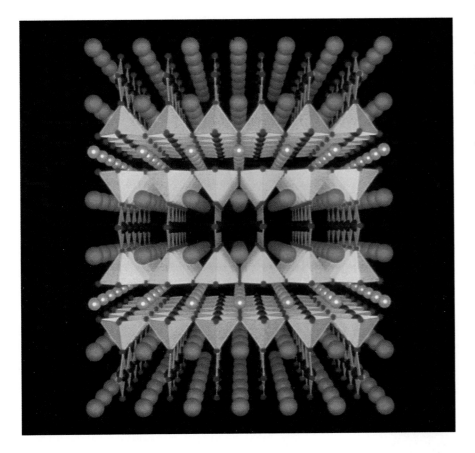

Two horizontal ceramic layers (centered on the planes of yttrium atoms) are sandwiched between three metallic layers in the superconducting ceramic whose chemical formula is YBa₂Cu₃O₇. This substance is called a 1-2-3 superconductor because of the relative abundances of the atoms of yttrium (gray), barium (green), and copper (blue). Oxygen atoms are colored red. The basic copper-oxygen structure illustrated on the right is repeated many times along the top, middle, and lower metallic layers.

shows the onset of superconductivity at about 125 K. These discoveries have inspired the unmet ultimate goal of finding a substance that is superconducting at room temperature (approximately 300 K), where no cooling would be necessary.

The electrical properties of the 1-2-3 compounds depend critically on the amount of oxygen that is chemically bound into the ceramic. For instance, if there are six oxygens instead of seven, the material is an insulator. The material becomes superconducting when the oxygen concentration reaches about 6.4, but the transition temperature is low. As more oxygen is added, the critical temperature rises until it reaches 93 K when the concentration is 7. Superconductivity similarly depends on composition in other types of compounds. The crystalline structure of many of these high-temperature superconductors has the common property that it consists of alternating layers of metals and insulators.

A computer-generated image of the structure of the 1-2-3 superconductor, determined from neutron scattering experiments, is shown on this page. Thanks to supercomputer simulations, as described below, and to experiments, we know that the conduction of electricity takes place in the layers that are arrangements of copper and oxygen atoms, seen along the top, middle, and bottom of the illustration. The arrangement of copper and oxygen atoms in two-dimensional planes seems crucial to their superconducting abilities.

To understand high-T_c superconductivity, scientists first need to know how the valence electron density is distributed for each energy level—that is, what the energy band structure is—when the material is in the normal, non-superconducting state. Art Freeman and his colleagues have used the FLAPW method described in Chapter 3 on Cray-2 supercomputers at NASA Ames Research Center and the University of Minnesota to investigate the ground-state electronic structure of the 1-2-3 superconductor. The FLAPW method yielded a charge density map, shown below, that displays the charge distribution for electrons with energies near the Fermi surface in two perpendicular vertical planes through the lattice shown schematically on this page. The electron density concentrations clearly show the copper-oxygen chains and planes. The supercomputer results tell us that the yttrium and barium atoms in the compound act as electron donors, but that they are above and below the plane of conduction, not in it.

The excitations that are responsible for the high-T_c superconductivity presumably occur near the Fermi energy surface, which is why it is so crucial to determine its details. To visualize the complex shape of the Fermi surface, it is customary to think of it in terms of the electron momentum, since energy can be thought of as the square of momentum (for example, kinetic energy is proportional to the square of velocity). One can plot the x, y, and z components of each conduction electron's momentum as derived from the supercomputer computations, producing an image like the one on the facing page from the work of W. E. Pickett, H. Krakauer, R. E. Cohen, and D. J. Singh. The image was obtained using local density functional methods on the NCSA Cray-2 and the Cornell Theory Center IBM 3090V supercomputers. Many precise experiments performed on these materials now confirm the general shape of the Fermi surface predicted by the computations both of Freeman and his colleagues and of Prickett, Krakauer, and Cohen.

The local density functional calculations described above tell us a great deal about the electronic structure of the 1-2-3 superconductor in its normal conducting state, but less in its superconducting state. The subtle excitations that transform this electron landscape into a superconductor are created by the collective interactions of many electrons and ions throughout the lattice. Since local density functional theory is a single electron approximation, it cannot easily compute these excited mechanisms. Nonetheless, calculations strongly indicate that some process other than the normal BCS electron-phonon interaction is required to explain the high values of T_c observed in these remarkable materials.

In recent years, researchers have proposed many theories to explain how these high-T_c superconductors work. However, given the complexity of the material and the growing precision

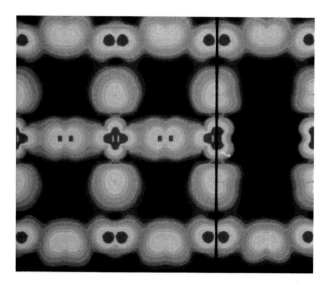

This color-coded map produced by Art Freeman and his colleagues shows the spatial distribution of the charge density near the Fermi surface in YBa$_2$Cu$_3$O$_7$ along two perpendicular planes. To the left of the vertical black line is a map of the z-y plane; the x-z plane is at the right. Note that the central copper atom is sandwiched between oxygen atoms in the y and z directions, but not along the x direction. Colors range from red indicating a high charge density through the colors of the rainbow to blue for a low charge density.

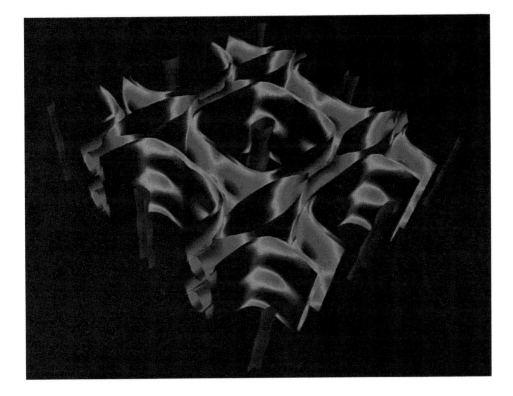

The Fermi surfaces of $YBa_2Cu_3O_7$, slightly broadened and extended periodically, as computed by Pickett, Krakauer, Cohen, and Singh. The colored surfaces show the values of the x, y, and z components of the momentum of the conduction electrons.

of the wide variety of experimental probes, the final explanation will probably only be attained after a supercomputer has derived detailed predictions from the theorist's general concept that can be compared with the experimental results. Recent work of Philippe Monthoux, Alexander Balatsky, and David Pines of the University of Illinois at Urbana-Champaign illustrates the contribution to be made by supercomputing.

They believe that interactions between the spins of electrons, rather than the electron-phonon interaction of the BCS theory, lead to the pairing of electrons. Their theory is based on the two-dimensional nature of the conducting copper-oxygen planes. Even though the physics is only two-dimensional, the strong electron-spin interaction requires the computation of a very complicated equation, in contrast to the weak electron-phonon interaction. Monthoux, Balatsky, and Pines have used the Cray Y-MP

supercomputer at NCSA to evaluate the model, which at this preliminary stage agrees with a number of experiments. It is still too early to say whether their theory or one of the many others will ultimately explain the high-T_c phenomena. What seems clear, though, is that for one of the first times in a debate among theorists supercomputers will play a crucial role in determining the outcome of the debate.

Buckyballs

Nearly as exciting as the discovery of the ceramic copper oxides was the recent discovery of a molecule composed of 60 carbon atoms. These carbon atoms occupy the vertices of 20 hexagons and 12 pentagons arranged like a geodesic dome. Called buckminsterfullerene, or "buckyball" for short, this molecule was discovered in 1985 and named by *Science* magazine as "Molecule of the Year for 1991." It was surprising that carbon, which has been so well studied in its two other structural forms of graphite and diamond, should remain so long undiscovered in this other, highly elegant form. Its hollow shape suggests its use as a catalytic agent that could trap smaller molecules inside its cage. Many coated buckyballs rolling over each other should make an excellent lubricant as well.

In solid form, pure buckyballs act as semiconductors. However, as small amounts of impurities are added (such as potassium, rubidium, cesium, or thallium), the doped buckyballs can become superconductors. For instance, as potassium is added, the semiconducting pure crystal changes to a conducting metal; at a composition of K_3C_{60} it becomes superconducting; and finally at a composition of K_6C_{60} it becomes an insulator. This behavior is similar to the behavior observed as oxygen is added to the 1-2-3 supercomputer. These doped buckyballs maintain superconductivity up to 45 K, almost rivaling the copper oxides as high-temperature superconductors.

Naturally enough, scientists studying potential supercomputing materials attacked the molecule immediately. The image on page 86 shows the electron density as computed by Jerry Bernholc, a professor at North Carolina State University, and his colleagues on the Cray Y-MP at the North Carolina Supercomputing Center. Using Car-Parrinello techniques to solve the local density functional equations, Bernholc and his colleagues found that the predicted electronic structure had some unusual properties, which were also observed experimentally. When the solid is heated, the energy supplied by the heat is transformed into local rotational energy, causing each buckyball to spin at a rate of billions of times per second! The buckyball structure remains intact at temperatures as high as 2000 K.

Art Freeman and his colleagues have used a modified version of the FLAPW method to investigate the superconductivity of the doped versions of this unique molecule. They find that the electronic structure determined by the local density functional method, when used in a theoretical formula to predict the transition temperature, seems to indicate that the BCS mechanism may be able to explain the superconductivity. Whether this new form of carbon can be manufactured in a practical form such as a wire remains to be seen, but its discovery certainly suggests that there are exciting new forms of material awaiting detection in the future.

Faster and Faster Switches

We move now from conducting materials to semiconductors, whose use in transistors has made the modern era of computing possible. This basic switch has developed a great deal since its invention in 1948. Whereas the original Bell Labs transistor had a switching time of microseconds, those in today's supercomputers can switch in a few nanoseconds. The unrelenting push to design ever smaller and faster transistors explains the continual increase in supercomputer speed during the past decades. Today, supercomputer simulations of novel electron devices are helping to ensure that speed continues to increase in the next generation of transistors.

Most transistors found in supercomputers today are made of silicon, an element from group IV of the periodic table. Silicon is chemically "doped" with small amounts of other elements, such as phosphorus and arsenic from group V, which easily donate electrons to the conducting state (*n*-type electron donors), or boron, aluminum, gallium, and indium from group III, which can accept electrons (*p*-type electron acceptors). These latter impurities can be viewed as providing "holes" that can move from atom to atom as if they were positive charges. Thus, a transistor can be viewed as a dilute system of charge carriers, both electrons and holes, moving through a solid.

Faster switching times can be achieved by building semiconductor devices directly out of a mix of elements from groups III and V. The most studied of these semiconductors is gallium arsenide, the semiconductor used in the Convex C3 supercomputer. Electrons in gallium arsenide have several times the mobility of those in silicon; the substance uses less power and requires less cooling than silicon; and it can emit light as well, and thus allows photonic and electronic functions to be combined in the same device.

To increase electron mobility even further, designers have created "compound semiconductors" called heterostructures, formed of sandwiches of different III-V compounds deposited in layers a few atoms thick. A particular example is the High Electron Mobility Transistor (HEMT), which is a heterostructure made of gallium arsenide and aluminum gallium arsenide. This device has a switching time of a few picoseconds, or a few one-trillionths of a second. Thus it switches a thousand times faster than the clock cycles of modern microprocessors!

The advantage of the heterostructure design is that the aluminum gallium arsenide layer can contain all the dopant atoms that donate electrons to the conduction bands in the gallium arsenide layer. After donating an electron, an atom becomes a positive ion that will scatter other electrons, slowing the current flow and creating electrical resistance. By placing the donor atoms and conduction band in separate layers, the designer minimizes the scattering of the mobile electrons.

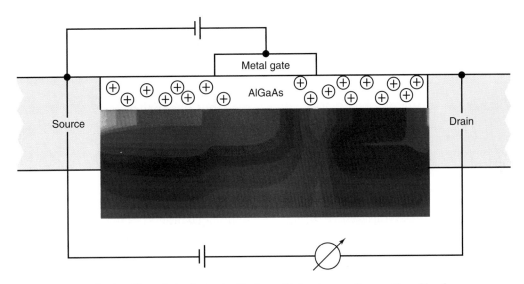

The source (or "emitter") and the drain (or "collector") for electrons lie on either side of a heterostructure made of aluminum gallium arsenide (AlGaAs) and gallium arsenide. When the voltage on the metal gate is switched from negative to positive, electrons begin flowing in a thin sheet below the AlGaAs layer, establishing a current that flows from the source material into the region under the gate in only a few picoseconds. The colors show the computed mobile electron density, which is highest if yellow, intermediate if red, and lowest if blue. The dimensions are set by the width of the gate, which is one micrometer.

An early probe of the switching characteristics of a HEMT device was carried out by Karl Hess, of the University of Illinois at Urbana-Champaign, and his colleagues in 1985 using a Cray X-MP located at Cray Research in Minneapolis. Hess and his colleagues followed the rapid diffusion of the mobile electrons through the semiconductor as the electric field is altered. Hess and his colleagues discovered from their simulation that it takes twice as long to switch on the device as to switch it off, because the electron velocities are significantly higher in the switching-off cycle. This asymmetry in switching time had been noticed experimentally, but was not completely understood. Since his experiment, Hess has developed a complete engineering design tool, Minilase, that simulates both the electronic and photonic properties of three-dimensional III-V heterostructures.

The gates in modern transistors are smaller than one micrometer in diameter, less than the length of a bacterium. Yet as small as these gates are, they are still 10,000 atoms across, much wider than necessary to carry the smallest conceivable current of a single line of electrons. It would seem that the quantum of the information world, the bit, is destined ultimately to be stored by the quantum of the physical world, the electron. Although it may seem like science fiction to write about single-electron transistors, the basic ideas have been around for decades and laboratories throughout the world are rapidly transforming the ideas into practical devices.

One of the most visionary of twentieth-century scientists, Richard Feynman of the California Institute of Technology, explored the limits placed by the laws of quantum physics on the tininess of mechanical or electronic devices in a 1959 talk he gave to the American Physical Society entitled "There's Plenty of Room at the Bottom." Assuming a redundant storage scheme of 100 atoms per bit of information, he concluded that, in principle, all the words in all the books in the world could be stored in a volume

as small as a dust mote one sees twinkling in the sunlight!

Feynman's ideas were decades ahead of their time. But by the 1980s, scientists were expressing great interest in new "micromachines" that could be created in the nanoworld, where objects begin to approach the smallness of atoms. The term "nano" is used because the clear limit to these schemes lies at a scale of about one nanometer, or one-billionth of a meter. Atoms are typically about 0.1 nanometer in diameter. Large protein molecules can be tens of nanometers in scale, viruses a few hundred nanometers, and animal cells 10,000 nanometers across.

As a device shrinks from 1 micrometer to 0.1 micrometer (or 100 nanometers), quantum effects become increasingly important. Thus scientists can no longer rely on the semiclassical approximation that suffices for simulating the electron behavior of present-day transistors. For instance, the wavelike nature of the electron becomes important as scales shrink to several wavelengths in dimension, and the macroscopic notions of resistance, capacitance, and induction yield to new and exotic microscopic ones. Many researchers believe that the family of technologies that has made the exponential shrinking of integrated circuits possible will reach its limits around the turn of the century as the features of electronic devices approach a few tenths of a micrometer in size. Thus scientists looking for a new technology to succeed the present one are rapidly accumulating a body of basic research on nanoelectronic systems, which are envisioned to exist on scales ten times smaller and which perform their electronic functions using purely quantum mechanical effects.

Because the wave nature of the electron begins to dominate at scales below 10 nanometers, one could use wave interference at these scales to create extremely small and fast switches. The idealized T-shaped device simulated on the facing page is a nanodevice about 100 nanometers long by 10 nanometers wide.

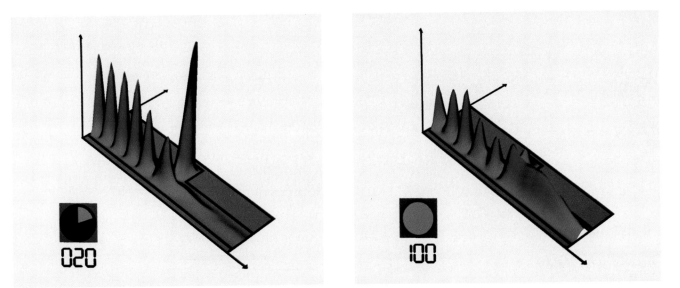

Destructive interference by an electron wavefunction in the stub halts the progress of electrons coming down the "quantum wire" (left), in this simulation by Umberto Ravaioli and L. Frank Register. A change in voltage moves the stub electron wavefunction by one-half a wavelength, allowing the flow of electrons to proceed (right). The standing waves in the stationary electron have individual peaks and troughs, while the moving electron wavefunction is continuous. The "clock" shows elapsed time in femtoseconds (thousandths of a picosecond).

The width of an electron wavefunction just fills the stub branching off the "quantum wire." The effective length of the stub is determined not by its material dimensions, but by the value of the voltage applied to the stub gate. As electrons come down the quantum wire, they sense the electron wavefunction in the "quantum dot" in the stub. At a long effective stub length, the wavefunction creates destructive interference that acts as a barrier to halt the movement of electrons. By suddenly changing the length of the stub by half, for example, one changes the interference from destructive to constructive, and the electrons continue their movement down the quantum wire. The "length" of the stub can be readjusted in the incredible time of a few tenths of a picosecond. Although this avenue of research is promising, the effect is very delicate and can be easily swamped by thermal fluctuations and impurities in the material.

Quantum electronic devices based on another quantum effect called tunneling are likely to be more robust. According to Heisenberg's uncertainty principle, the wave nature of matter intrinsic to quantum theory imparts an inherent "fuzziness" to the location of subatomic particles. Thus an individual electron can occasionally be found some distance from its most probable position. At such times, the electron will appear to have spontaneously jumped from one location to another without actually traversing the intervening space. If there is an energy barrier between an electron's most probable location and the site where the electron is actually observed, then physicists say that the electron has "tunneled through" that barrier.

This tunneling property can be used to pass one electron at a time through an insulating barrier 5 to 10 nanometers in width called a tunnel junction. If two of these tunnel junctions

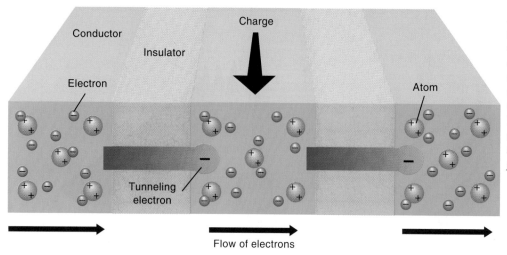

Conductor

Insulator

Charge

Electron

Atom

Tunneling electron

Flow of electrons

By changing the charge on the middle electrode by only one-half the charge on an electron, one can induce a single electron to tunnel through the insulating region from the source to the drain. Transistors that exploit tunneling should be able to switch many billions of times per second.

share a middle electrode as in the diagram on this page, they form a single-electron transistor. The application of a voltage to the electrode causes a buildup of electric charge that lures electrons across the insulator, creating a current. Experimental versions of this device, such as those constructed by Konstantin Likharev and his colleagues at the State University of New York at Stony Brook, are 60 to 100 nanometers in size, but there seems to be no reason such a device could not be shrunk to a length of 10 nanometers. As the size of these tunneling devices begins to approach a few nanometers, supercomputer simulations that fully consider quantum effects will become necessary to fine-tune their design.

At these tiny scales, a square centimeter of semiconducting material could comfortably hold well over 10 billion transistors, a thousand times today's highest densities! Operating at near absolute zero temperature, each single-electron transistor would in principle perform the function of a relay or vacuum tube of fifty years ago, but it would be ten to a hundred million times smaller and faster. Coupled together in a massively parallel architecture, these or other nanoworld devices should keep the speed of supercomputers progressing at exponential rates for most of our lifetimes.

Moving Atoms Around

A new generation of laboratory instruments has been created for viewing and manipulating materials at an atomic level of resolution. In the early 1980s, Gerd Binning and Heinrich Rohrer at the IBM Research Laboratory in Zürich, Switzerland, invented the scanning tunneling microscope (STM), for which they received the Nobel prize in 1986. Since then other instruments have been developed, such as the atomic force microscope and new nanolithography technologies, creating a rich kit of tools for investigating the nanoworld.

A scanning tunneling microscope makes use of the tunneling phenomenon discussed above. The device contains a very sharp metal electrode, which is brought within a few atomic diameters of the surface to be examined. A small voltage, low enough that no current would normally flow across the gap, is applied between the probe and the surface. Because of the wave nature of subatomic particles, however, some electrons from the electrode do in fact tunnel across the gap, resulting in a measurable "tunneling current." The exploitable feature of this current is that it varies exponentially with the

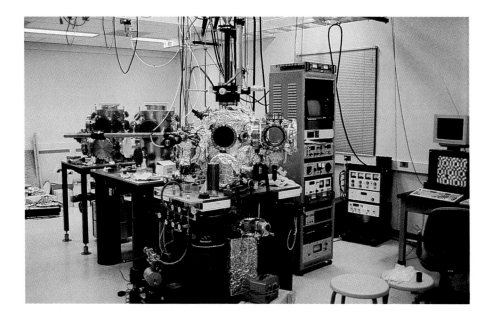

A modern ultra-high-vacuum STM lab, such as this one in the University of Illinois Beckman Institute, enables researchers to explore and modify surfaces at the atomic level. The individual specimens to scan are placed in the vacuum chambers on the table, and the location of the scanning tip is remotely controlled by computer. The image appears (here the surface of silicon) on a graphics workstation.

distance between the probe and the surface. For instance, a change of only one atomic diameter in the gap's width can produce a tenfold change in the current.

The tunneling current is therefore an extremely sensitive measure of the distance between the tip of the probe and surface. To explore the topography of the face of a material, the probe is moved from side to side across the surface. As the probe scans the surface, the tunneling current is modulated by the distance between the one or two atoms at the very tip of the probe and the atom on the surface nearest the probe. By controlling the probe height to maintain a constant tunneling current, a map can be constructed that shows the arrangement of atoms on the surface.

As it scans a surface, an STM produces an enormous stream of data. The most recent STMs can take one million height measurements in a few seconds. NCSA staff members Clint Potter and Rachael Brady, working with the Beckman Institute STM laboratory of Joseph Lyding and John Tucker, have connected a Convex C3 supercomputer to the STM so that they can process this information almost instantly, producing an image of the surface as the scanning proceeds. The group has developed software that allows researchers to zoom in and survey a much smaller area at atomic resolution when they notice something of interest in a wide-field view. Using a joystick coupled to a graphics workstation connected both to the STM and supercomputer, the researcher can "fly" over the surface and view its atoms on a color monitor.

An STM tip is also able to modify surfaces by manipulating single atoms. IBM researchers caught the world's attention in 1990 when they moved individual xenon atoms on a nickel surface, lining the atoms up to spell "IBM." A variety of methods under development will enable STMs to deposit layers of atoms or to act as extremely small "drills." The STM image on the next page shows a graphite surface after the Lyding and Tucker group used high-electric-field sputtering to create three one-nanometer-wide holes. One can count the individual carbon atoms and determine that the holes are only five atoms wide!

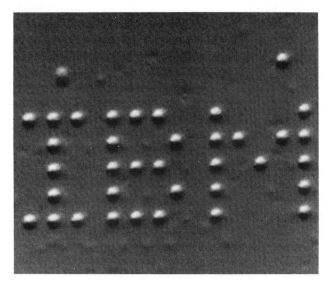

Single xenon atoms have been arranged on a nickel surface through the use of a scanning tunneling microscope to spell out "IBM."

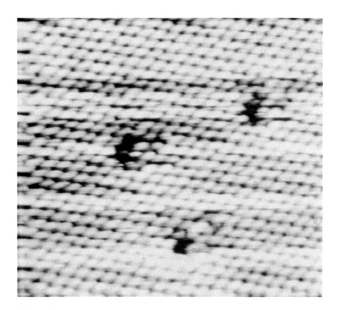

When Lyding and Tucker's group applied short pulses of high voltage to the STM tip in a high-vacuum chamber, the ion sputtering created holes a few atoms across in the graphite surface. The instant visual feedback supplied by a networked computer will make this technique a practical tool for modifying surfaces.

Supercomputers can be used to simulate the atom-by-atom modification of a surface by an STM tip. On the facing page are the intriguing results of a molecular dynamics simulation carried out by Uzi Landman and W. D. Luedtke at the Georgia Institute of Technology using a Cray Research supercomputer at the National Magnetic Fusion Energy Computing Center in Livermore, California. This sequence of images shows what happens as an ultrafine nickel tip (about ten times smaller than the STM tip used above) contacts and then pulls up from a gold surface. The interatomic forces form a monolayer of gold atoms that sticks to the nickel tip. A connective neck of atoms from three layers deep in the gold builds up as the nickel tip is withdrawn. As refined versions of such simulations become a standard tool of nanotechnology in the future, scientists will be able to plan in advance the most effective manner of modifying a surface on the atomic level.

Surface Catalysis

We move finally to the world of chemistry, where supercomputers are used to compute the electronic structure of individual atoms and molecules.

Our first example, the attachment of a molecule to a surface, is halfway between modeling the bulk characteristics of a semiconductor, say, and modeling individual reacting molecules. The interactions of molecules on a surface are important in the electrochemical processes used to chrome plate car bumpers and produce electricity in batteries. At the heart of electrochemistry is the interaction of water with the surfaces of metal electrodes. To chrome plate a car bumper, for example, a current is sent through the bumper, which functions as an electrode. The current flows from the bumper through an intervening electrolyte solution of ions and water to an anode. The electric current removes electrons from chromium ions in solution; as a result, chromium metal is deposited on the

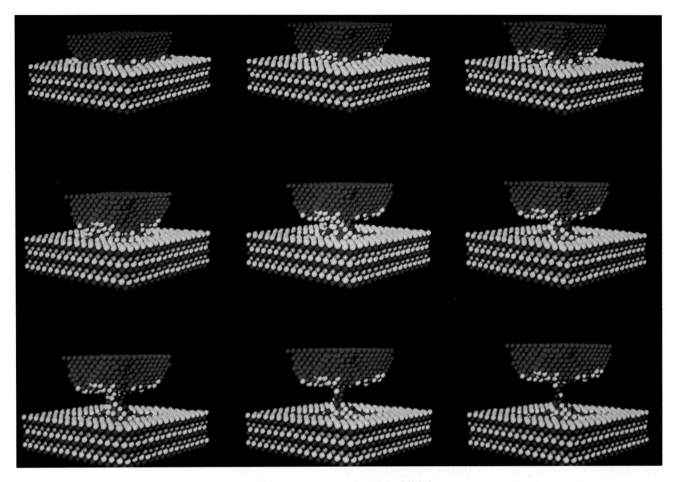

An ultrafine nickel tip pulls up gold atoms as it withdraws from a gold surface. Nickel atoms are colored red; gold atoms in the top layer are yellow, in the second layer blue, in the third layer green, and then yellow again.

bumper surface. The electrons are carried by water molecules to the anode, where they are given up to the surface atoms to complete the circuit. One of the most widely used electrodes is the mercury falling-drop electrode, which continuously provides a clean surface to the water.

One can imagine that the molecular dynamics of thousands of water molecules interacting with the metal surfaces and other key molecules in the electrolyte solution would be a powerful computational tool to help in the un-

derstanding and designing of electrochemical reactions. However, as we saw in Chapter 3, molecular dynamics is only as good as the force fields or potential energies from which they are derived. Therefore, the developers of molecular dynamics codes need a high-quality potential energy function describing the interaction of water with the metal surface.

Harrell Sellers and Pamidighantam Sudhakar, working at the NCSA, have very recently performed accurate quantum chemistry computations of the electronic structure a water molecule binding to a mercury surface. They used *ab initio* Hartree-Fock methods to compute the electron distribution, correcting for the correlation effects with perturbation theory. Because mercury has tightly bound inner electrons, special relativity corrections were also important to obtain an accurate answer. The computations were carried out on an IBM RISC workstation and on the NCSA's Cray-2 supercomputer.

Sellers and Sudhakar modeled the mercury surface as two layers of mercury atoms. The outer 12 electrons of each of the 7 close-packed atoms in the upper layer were computed quantum mechanically as they interacted with each other and with the electrons in the water molecule. The lower 15 atoms were represented by fixed potential energies, which did not change as the water approached the surface. The distances between the mercury atoms were held fixed at separations derived from bulk measurements. Similarly, the geometry of the water molecule was held rigid, but its distance and orientation to the surface were varied.

They found that the water molecule positions itself vertically so that the hydrogen atoms are farthest from the metal surface. Because the single hydrogen electron spends part of its time orbiting the oxygen atom, the hydrogen side of the molecule is electrically slightly more positive, and correspondingly the oxygen side is slightly more negative. As the water molecule approaches the mercury surface, the negatively charged oxygen repels electrons in the mercury atoms. The anode surface becomes slightly more positive, creating a weak attraction between the oxygen and mercury atoms. Sellers and Sudhakar also detect a certain amount of mixing between the electron orbits of the mercury and oxygen. By calculating the electron distribution for the water molecule at about 40 different positions over the mercury surface, they are able to map the potential energy for this interaction at the high accuracy required.

The Chemistry of the Ozone Layer

Designing a new molecule involves the same strategy as designing a new material: the chemist must calculate the arrangement and positions of all the atoms in the molecule. This is accom-

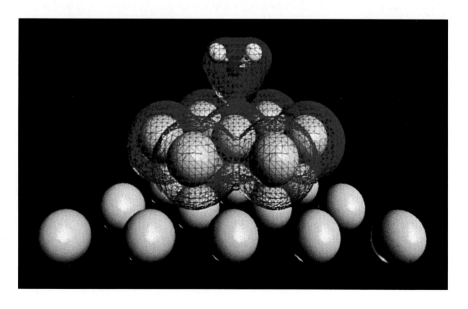

The quantum chemistry computation of the electronic structure of a water molecule near the surface of a mercury electrode. The red mesh surrounds the volume inside of which 90 percent of the electron density is located. The sphere diameters are proportional to the atomic radius, with the hydrogen atoms in white, the mercury atoms a metallic silver, and the oxygen atom in blue.

plished by solving the Schrödinger equation, which in this case yields the distribution of electrons around the nuclei of the atoms of a molecule. By solving for the electron configuration of lowest energy, the chemist obtains not only the geometry of the molecule, but also a description of its vibrational modes and its reactivity.

The vibrational modes of a molecule tell us how that substance interacts with infrared radiation: a molecule can switch from one quantized vibrational mode to another by absorbing or emitting an infrared photon that equals the difference in energy between the two modes. The strength of absorption in the infrared is a crucial property of substances that go into the atmosphere, because the driving force behind the greenhouse effect is the absorption of infrared radiation by atmospheric gases such as carbon dioxide, methane, and the chlorofluorocarbons (CFCs).

The last of these greenhouse chemicals is also implicated in the depletion of the Earth's protective ozone layer. Chlorofluorocarbons are nonreactive until they reach the stratosphere, where the Sun's ultraviolet light breaks the chlorine bonds in these molecules, releasing reactive chlorine. Small particles such as ice crystals in the high-atmosphere clouds act as surfaces on which the chlorine can catalyze the destruction of ozone. During its lifetime in the stratosphere, one chlorine atom can break up as many as 100,000 ozone molecules. As a possible result of the destructive effects of CFCs, the recurring hole in the ozone layer over the Antarctic has recently been joined by one over the high latitudes in the Northern Hemisphere. Acknowledging the seriousness of the situation, most countries in the world have signed the Montreal protocols, which call for production of CFCs to be phased out by 1995.

CFCs are currently used throughout the world as coolants in refrigerators and air conditioners, as well as in cleaning fluids and aerosols; thus an effective chemical substitute must be found before the ban goes into effect. The new molecule must not only be a good cooling agent with little toxicity, but must also have low infrared absorption to avoid enhancing the greenhouse effect and a short atmospheric lifetime.

A new class of compounds seems to hold great promise: these are the hydrofluorocarbons (HFCs), formed by replacing the chlorine with hydrogen. They possess both environmentally damaging qualities to a much lesser extent than CFCs, because the presence of the hydrogen causes these compounds to be much more reactive. As a consequence, they will decompose in the lower atmosphere by combining with hydroxyl radicals and therefore will not survive long enough to reach the stratosphere where they can attack the ozone layer. As of mid-1992, the Environmental Protection Agency has selected nine of these compounds for further study.

With a supercomputer, a chemist working with atmospheric scientists can assess the potential impact of a substance on the atmosphere before it is mass produced. David Dixon and his colleagues at Du Pont have been using a supercomputer for just this purpose to help create commercial alternatives to the CFCs. Dixon attacked three problems that must be resolved if these compounds are to be produced in high volume by mid-decade. First, Hartree-Fock-based methods were used to determine the geometry of the currently used CFCs and their potential replacements. Second, these geometries became starting points for accurate calculations to predict the reaction rates of the new molecules with the hydroxyl radical. A computation using the code Gaussian 90 on the Du Pont Cray Y-MP supercomputer, similar to that yielding the vibrational modes, provides this quantity. Third, Dixon is working to develop a new form of industrial catalysis that can produce these more reactive molecules in bulk. He used local density functional methods on this problem, but the computation is quite involved because of the complex electron structure of the most promising metal catalysts.

The detailed information calculated for the new molecules together with experimental data was inserted into Du Pont atmospheric chemis-

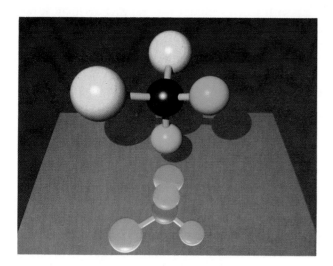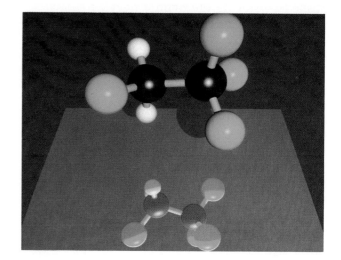

The optimized geometry, computed by quantum chemistry techniques, of one of the most widely used chlorofluorocarbons, CFC-12, and a potential replacement, HFC-134a. Carbon atoms are black, fluorines are green, chlorines are yellow, and hydrogens are white. To aid in visualizing the three-dimensional structure, both shadows and a mirrored surface are shown under each molecule. Computations and visualizations by David Dixon of E. I. Du Pont de Nemours & Company and Pat Capobianco of Cray Research.

try codes to predict the impact of these new molecules on global warming and ozone depletion. The results show the proposed substitutes to be much more environmentally benign. The multibillion dollar a year CFC market must have a replacement in three years, so time is precious. Dixon estimates that it would take roughly three months at a cost of $50,000 for the National Institute of Standards and Technology to measure experimentally the properties of one new compound, once a large enough quantity has been synthesized to measure. He can compute values for these properties in a few days for under $5000 on Du Pont's Cray Y-MP supercomputer. Furthermore, the accumulating database on CFCs and their replacements will be a valuable advantage for his company as the race to develop a CFC replacement intensifies.

Academic researchers are using supercomputers to solve the quantum mechanical equa-tions that define ozone-destroying chemical reactions. For instance, Almon Turner of the University of Detroit has been using NCSA's Cray systems to study couplings between three major cycles of chemical reactions, all leading to ozone depletion. In addition to chlorine, two other families of compounds, nitrogen oxides and hydrogen oxides, participate in catalytic cycles that break up ozone molecules. Although these compounds are typically assumed to act independently in the stratosphere, Turner is concerned that there may be couplings between the cycles, since air samples collected in the stratosphere have shown the existence of such molecules as chlorine nitrate, $ClONO_2$, and per-nitric acid, HO_2NO_2.

By solving the Schrödinger equation, using Hartree-Fock methods, Turner is able to com-pute the reaction rates between the major mole-cules in the three reaction cycles and thus pre-

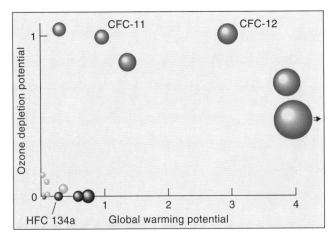

Du Pont researchers Donald Fisher, Charles Hales, Asron Owens, and David Filkin used atmospheric chemistry codes to compute the ozone depletion potential and the global warming potential for the currently used CFCs (royal blue) and the molecules proposed to replace them, hydrofluorocarbons (purple) and hydrochlorofluorocarbons (turquoise).

dict which couplings are the most important. This basic research may be important in formulating future environmental policy, as a more comprehensive picture develops of the mechanisms creating this major environmental crisis.

The Mating Rites of Molecules

Although the atmosphere contains gases of many kinds, their low density allows two reacting molecules to interact as though in isolation from all others. In contrast, if the reactants are dissolved in a solvent like water, then even though the water molecules may not participate in the reaction per se, their presence can significantly alter the distribution of electrons around the reacting species. Such interference can cause the chemical reaction to differ in important ways from the same reaction in the gaseous phase.

Using supercomputers, chemists can simulate the effect of hundreds to thousands of water molecules surrounding two reacting species. Unfortunately, today's supercomputers are still not powerful enough to treat the large number of atoms involved purely quantum mechanically. However, by cleverly combining techniques of quantum chemistry and molecular dynamics, chemists can make the problem tractable.

In early 1988, Paul Bash, Martin Field, and Martin Karplus from Harvard University collaborated with NCSA visualization staff Matthew Arrott, Michelle Mercer, and Jeffrey Yost to visualize a chemical reaction in the presence of the water molecules. As a test case they chose the simple exchange reaction between a chlorine ion (Cl^-) and a methyl chloride (CH_3Cl) molecule. Methyl chloride, which is used as a refrigerant and a local anesthetic, consists of a chlorine atom attached to a methyl group composed of one central carbon atom surrounded by three hydrogen atoms. In the reaction studied by Bash and his colleagues, the methyl group swaps its chlorine atom for the chlorine ion, which becomes attached to the side of the methyl group opposite from the original chlorine atom.

The Schrödinger equation was solved on NCSA's Cray X-MP using quantum chemistry Hartree-Fock techniques to determine the electronic structure of the methyl chloride, while the motion of the water molecules was handled using molecular dynamics methods. Classical methods adequately describe the effects of the water molecules because the water molecules affect the reaction primarily through the electric fields that surround their hydrogen and oxygen atoms rather than by a quantum process of exchanging electrons, unlike the case in Sellers' work discussed earlier. The effects of their electric fields were directly inserted into the quantum mechanical computations, which then produced the wavefunctions for the chlorine ion and the methyl chloride molecule.

From experiments it is known that water has a significant effect on this reaction. In the visualization, one can clearly see why. In a simulation run without water molecules, the chlo-

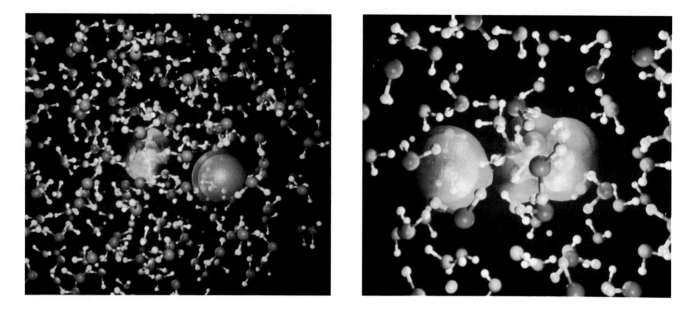

Two moments in the chemical reaction of a chlorine ion and a methyl chloride molecule in an aqueous solution. The simulation tracked the motions of 180 water molecules along with the reactants in a sphere 1.1 nanometers across. The charge density around the reacting species was determined from the wavefunction, and displayed in two levels of density shown by the yellow and green surfaces.

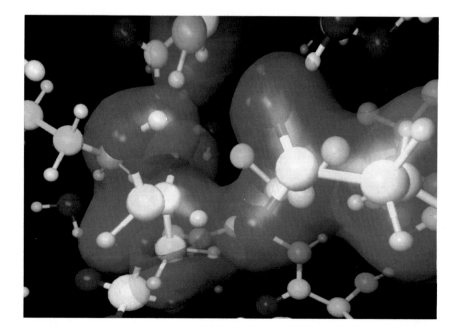

A white hydrogen atom is shown in mid flight between a green carbon and a red oxygen atom, as a protein, on the right, exchanges hydrogen atoms with another molecule. The purple surface represents a constant value of electron density. Note the water molecules formed of one red and two white atoms. The yellow sphere is a phosphorus atom.

rine ion simply moves toward the methyl chloride in a straight line and forms a bond. However, when numerous water molecules are present, they shield the chlorine ion. The hydrogen atoms in water molecules carry a slight positive charge, and so they are attracted toward the negative charge on the chlorine ion, which they tend to surround.

In order for the reaction to proceed, some of these water molecules must be removed. Since removing the water molecules requires the addition of energy, it takes roughly three times the amount of energy to drive this reaction to completion as to drive the reaction in the gaseous phase. Furthermore, as the ion and the molecule move closer, the electron distributions develop bumps and dimples on the surfaces where the reactants face each other. When the simulation is performed in a vacuum, the electron clouds around the reactants are free of such irregularities.

Having developed these computational and visualization methods for the simple molecule methyl chloride, Bash and Karplus then teamed with Robert Davenport of MIT to study a more complex molecular interaction involving the transfer of a hydrogen atom between a small protein and a molecule it is docking with in an aqueous solution. Their simulation shows the transfer of additional atoms between the molecule and the protein as well, and the diffusion of the two molecules away from each other through the water after the transfers are completed. This sort of interaction between organic molecules is a good example of the type of activity that we will study at some length in the next chapter on the chemistry of life.

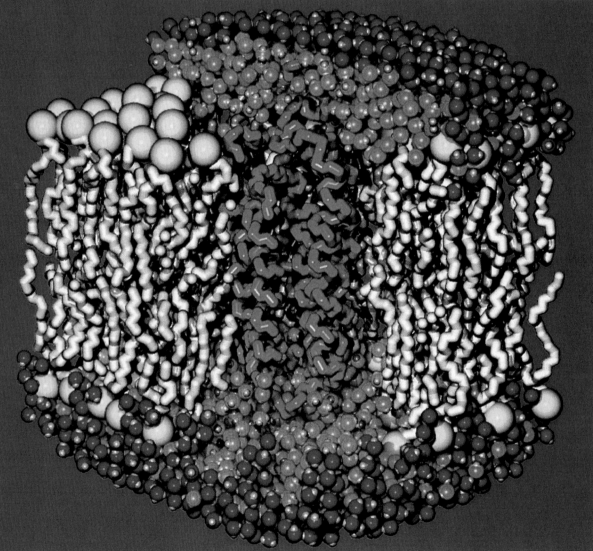

Inside Living Creatures

The Earth's millions of species range from microscopic bacteria to gigantic blue whales and towering sequoia trees. Yet in spite of their differences, all living things share a common chemical basis in the activities of two classes of biomolecules: the nucleic acids DNA and RNA, and the proteins. These molecules obey the laws of physics and chemistry and therefore can be modeled by the methods of quantum chemistry and molecular dynamics discussed in Chapter 3.

More profoundly, these molecules carry information. The linear sequence of the DNA molecule is effectively a master computer program, which by directing the production of proteins makes possible the enormous variety of living organisms. Proteins carry in their three-dimensional folded shapes the implicit information that

The bacteriorhodopsin photosynthetic protein (green) spans the cell membrane, as computed in a molecular dynamics simulation. The molecules of the lipid bilayer cell membrane molecules are oriented so that the tails (grey) face inward and the heads (yellow) line the surface. On either side of the membrane are water molecules, some free (orange) and others fixed in space (red).

allows them to dock with other molecules and so perform most of the biochemical functions of living creatures. We can use supercomputers to read out and analyze the information content of these molecules as well as to model their interactions. Furthermore, we can use these machines to elucidate how these and other molecules join together to form structures such as cell organelles or cell membranes, which in turn form the cells that build more complex structures, such as the circulatory system or the nervous system. These systems interact with each other to form the functioning organism.

In this chapter, we shall see that computing is beginning to make inroads into what has historically been a wet lab science. Supercomputers now allow researchers to perform vast numbers of "dry" numerical experiments at the molecular level. It seems inevitable that supercomputers will become indispensable to the biological and medical sciences, because of the need to decode the enormous information content of the genome, to understand the complex interactions of the large macromolecules, and to create models of the hierarchical subsystems of living organisms.

Genomes: The Computer Program of Life

Life on Earth is based on the ability of the molecular polymers ribonucleic acid (RNA) and deoxyribonucleic acid (DNA) to carry information. These two nucleic acids are long sequences of much smaller molecular structures called nucleotides, each of which is built from a phosphate group, a sugar, and a ringed structure called a base. In DNA the four possible bases are adenine (A), guanine (G), thymine (T), and cytosine (C), whereas in RNA, the thymine is replaced by uracil (U). Thus, one can use the order of the base letters to specify the order of the nucleotides.

The order of the nucleotides encodes instructions on how to build and maintain a living organism. A sequence of three nucleotides codes for one of 20 possible amino acids, the small molecules that are strung together in a chain of tens to thousands to form a protein. A gene consists of a stretch of DNA that defines the order of the amino acids in a given protein. The sequence of nucleotides in the DNA thereby stores the amino acid sequences for all the proteins in an organism.

The genetic representation of a protein, written in the DNA code, is not translated directly into that protein, however. First, a "working copy" of the gene is made in the form of an RNA molecule that carries the same nucleotide sequence. This so-called messenger RNA (mRNA) is the "tape" that feeds through a cellular "tape reading" machine called the ribosome, itself a complex of protein and special ribosomal RNAs (rRNAs). Molecules called transfer RNAs (tRNA) find the correct amino acid for each three-letter set of the mRNA code and bring it to the ribosome, depositing it on the end of the growing protein chain. Once the ribosome has released the completed protein, the chain of amino acids folds into a three-dimensional structure with a distinct shape. Its shape allows the protein to function as an enzyme, an antibody, a chemical communicator, a DNA regulator, or a structural element in a cell.

It would require three million pages this size to print out the sequence of three billion bases composing the human genome. This vast amount of information is packaged with remarkable compactness. The DNA molecule consists of two strands of nucleotides twisted into a helix measuring about 2 nanometers in diameter and 3.4 nanometers per twist along the helix. Since there are 10 nucleotides per twist, simple arithmetic reveals the length of a human being's DNA to be one meter; yet the supercoiled DNA that forms our chromosomes is contained within a cell nucleus extending only a few micrometers across. Our bodies store something like a gigabyte of information per cubic micrometer. Com-

```
                5          10         15          20          25          30         35
Human        N L V N F H R M I K – L T T G K E A A L S Y G F Y G C H C G V G G R G
Rattlesnake  S L V Q F E T L I M – K I A G R S G L L W Y S A Y G C Y C G W G G H G
Bovine       A L W Q F N G M I K C K I P S S E P L L D F N N Y G C Y C G L G G S G

                40         45         50          55          60         65        70
Human        S P K D A T D R C C V T H D C C Y K R L E K – R G C G – – – – – T K F
Rattlesnake  L P Q D A T D R C C F V H D C C Y G K A T – – – D C N – – – – – – P K T
Bovine       T P V D D L D R C C Q T H D N C Y K Q A K K L D S C K V L V D N P Y T

                75         80         85          90          95        100        105
Human        L S Y K F S N S G S R I T C – A K Q D S C R S Q L C E C D K A A T C
Rattlesnake  V S Y T Y S E E N G E I I C – G G D D P C G T Q I C E C D K A A A I C
Bovine       N N Y S Y S C S N N E I T C S S E N N A C E A F I C N C D R N A A I C

                110        115        120         125         130
Human        F A R N K T T Y N – K K Y Q Y Y S N K H C R G S T P R C
Rattlesnake  F R D N I P S Y D N K – Y W L F P P K D C R E E P E P C
Bovine       F S K – – V P Y N K E H K N L D K K – N C
```

The amino acid sequence for a protein called phospholipase A2 (PLA2), found in rattlesnake venom, shows a large amount of overlap with the amino acid sequence of the PLA2 protein in humans and cows. Each letter stands for one of the 20 amino acids.

pare their storage capacity with the information density of 10 megabytes per square centimeter for a modern solid-state electronic device!

After years of heated debate, the biological community has embarked on the Human Genome Project, whose goal is to chart the order of the nucleotides in the DNA of human cells and in a wide range of other organisms such as bacteria, yeast, worms, fruit flies, and mice. Because genes tend to be conserved over the course of evolution, 70 percent of a human's genes are identical with those of a mouse. In fact, the genes of the global human population are estimated to be identical to one part in a thousand. Thus, all our physical dissimilarities, including some 3000 genetically linked diseases, are contained in those small differences.

The comparison of genes and their expression both within and across species is forming a new basis of knowledge about biological systems and is creating new aproaches to medical treatment. In addition, this new-found genetic knowledge is already having an impact on our ability to modify traits of plants and animals. Genes determine the color of a tomato, the ability of grain crops to resist drought, the amount of milk a cow produces, or the ability of bacteria to digest oil. The information content of national gene data banks applied with the technologies of the laboratory will allow us to alter these and many other genetic traits.

As the rate of sequencing accelerates, national databases containing known sequence fragments are growing rapidly. As researchers apply ever more sophisticated software to discovering similar patterns among different genes, they find ever more often that they need a supercomputer. An early example was the use of a Thinking Machines Corporation's CM-2 in 1987 to compare every protein sequence with every other protein sequence in a national data bank of 2000 proteins, containing 730,000 amino acids. The comparison was accomplished in two hours on a CM-2, using a program developed by Eric Lander of MIT and Jill Mesirov and Washington Taylor IV of Thinking Machines Corporation. Their work evaluated various comparison algorithms and revealed many fragments of sequence in one protein that matched fairly well with other fragments in different proteins.

In 1992, Gaston Gonnet, Mark Cohen, and Steven Benner of the Swiss Federal Institute of Technology in Zürich, Switzerland, reported an exhaustive matching of all subsequences in a database containing over 8 million amino acids. By using a novel algorithm, they were able to make 35 billion comparisons in less than five months by taking advantage of whatever time was available on six DEC workstations running in parallel. Sequence matching is inherently suited to parallel processing and will be one of the first applications used on new architectures.

During the process of evolution, an entirely new genetic sequence is rarely invented, but rather older subunits are mixed and modified to achieve new functions in the proteins coded by the new genes. The Swiss-led database search and others like it should soon begin discovering the most ancient common units from which today's proteins are assembled. Turned around, this argument suggests that we may be able to predict some aspects of protein structure just from a knowledge of the gene's sequence. As we shall see shortly, this tactic is beginning to be exploited successfully by both university researchers and pharmaceutical companies.

The Tree of Life

The organism can make mistakes in copying the genetic sequence during cell reproduction. For instance, a nucleic acid base at a specific location may be switched with another base, leading to a letter change in the DNA code. Alternatively, a stretch of bases in a gene may become deleted or duplicated in a new location. If not lethal to the organism, mutations accumulate over evolutionary time, perhaps ultimately differentiating a new species. By comparing similar genetic sequences from different species, letter by letter, scientists arrive at a fractional number representing the nucleotide substitutions per total number of nucleotides. If they know how fast mutations accumulate, then they can calculate the length of time since two organisms began diverging from a common ancestor. The rate of mutation thus functions as a "molecular clock." Certain regions of the RNA or DNA are more susceptible to mutation than others, yielding molecular clocks of different rates, some for rapidly evolving sections of DNA that accumulates mutations over a matter of years (allowing us to track flu virus changes), some for more slowly evolving DNA that accumulates mutations over tens of thousands of years (allowing us to study the evolution of human beings), and some for very slowly evolving DNA that accumulates mutations over millions to billions of years (allowing us to study the evolution of all of life).

The rRNA of ribosomes makes a good long-term molecular clock. By using computers to compare the sequences of ribosomal RNA across many species, University of Illinois researchers Carl Woese, Gary Olson, and their colleagues have found the most probable evolutionary tree for the major groups of organisms. It shows that the species of life on Earth belong to three great kingdoms that have evolved separately for billions of years: the archaebacteria, the eubacteria, and the eukaryotes. Many archaebacteria thrive in primitive, hot, oxygen-deprived environments like deep-sea vents and the hot springs in Yellowstone, suggesting that they were among the earliest forms on Earth. The eubacteria include most common bacteria, which cause disease, decompose organic material, and fill test tubes in biomolecular laboratories. The eukaryotes include all the nonbacterial

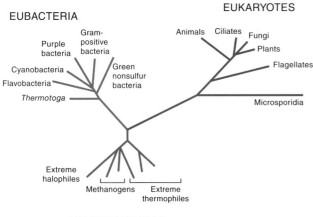

EUBACTERIA

Purple bacteria
Gram-positive bacteria
Cyanobacteria
Flavobacteria
Green nonsulfur bacteria
Thermotoga

EUKARYOTES

Animals Ciliates Fungi
Plants
Flagellates
Microsporidia

Extreme halophiles
Methanogens
Extreme thermophiles

ARCHAEBACTERIA

This evolutionary tree shows the most likely branching of the three great kingdoms of life as computed by comparing sequences of RNA in a large number of bacteria species and animals. The length of the branches is proportional to the number of mutations and therefore to the evolutionary age of tree branches.

life forms; they differ from bacteria in having cells whose genome is enclosed in a nuclear membrane.

For the first three and a half billion years of the evolution of life, there existed only the two kingdoms of bacteria and the one-celled eukaryotes. During the first billion and a half years, the Earth's atmosphere could support only anerobic life forms, which did not require oxygen for respiration. About two billion years ago, living matter completely altered the composition of the atmosphere by producing large amounts of free oxygen as a byproduct of photosynthesis. In turn, new metabolic processes evolved that made use of that oxygen. During the last billion years, eukaryotes underwent a burst of evolution that produced the first multicellular forms, which ultimately populated sea, land, and air. The history of the co-evolution of organisms and the Earth's environment is locked up in the genomes of living creatures.

As the genome-sequencing program continues, we will be able to expand this "genetic archaeology": from determining the genealogy of organisms we will proceed to determining the genealogy of the genes themselves. As mentioned above, there is great conservation of genes and their subcomponents across evolutionary history. Recently, Gary Olson and researchers at Argonne National Laboratory used the massively parallel Intel Delta supercomputer at Caltech to compare rRNA sequences from 473 microorganisms. This ground-breaking research uses algorithms that had been thought in the past to be too computationally intensive. They will allow scientists to hypothesize possible evolutionary paths for genes and then test these paths against the data embedded in the ensemble of gene fragments. As these techniques are refined, and the number of decoded gene fragments from many different species grows rapidly, our detailed understanding of how life and the Earth co-evolved will expand enormously.

The Protein Folding Problem

A linear string of amino acids folds into a complex three-dimensional shape, creating a functional protein. These structures can be long fibers, like the fibers of muscle, or they can have pockets and protuberances that fit tightly with other molecules, like a hand fits a glove. Examples of proteins with the latter type of structure are the antibodies that bind with molecules on the surface of invading pathogens. Proteins of the appropriate shape can come back and dock with the very DNA that gave them birth, in order to regulate the amounts of other proteins produced. Or, as enzymes, they can accelerate the rates of the thousands of biochemical reactions that make living creatures possible. At present we know the folded shape at the atomic level of resolution for about 500 proteins. Considering that the human genome codes for 100,000 proteins, it is clear than we will soon have orders of magnitude more genetic sequences for proteins than we will have three-dimensional shapes.

If each protein had a totally different shape, then the problem of determining protein structure would be hopeless. Fortunately, scientists have found that proteins belong to families of common evolutionary origin. Although there is a great deal of diversity within families, all family members share a common structural pattern. Thus, the common thread throughout the different approaches to determining protein structure is "pattern recognition." We shall illustrate the three major approaches to the determination of protein structure: x-ray crystallography, amino acid homology, and the application of novel ideas from other fields such as the study of neural networks.

Historically, x-ray crystallography has been the major technique used to determine protein structure directly, although scientists may now turn to a variety of new methods such as nuclear magnetic resonance or direct observation by atomic-level microscopy. X-ray crystallography was developed early in this century by Lawrence Bragg as a method of determining the lattice structure of inorganic crystals. Since x-rays have a wavelength of 0.1 to 0.2 nanometer, roughly the size of an atom, the waves will scatter or diffract off the electron distribution of the atoms of a crystallized protein, forming an array of diffraction spots on photographic film. From the details of the diffraction pattern, a computer can calculate a three-dimensional electron density map of the protein.

This map must then be interpreted in light of the known amino acid sequence for the protein, which forms the protein's "backbone." A tentative structure is proposed for the folded protein, and its atomic structure is compared with the electron density map inferred from the diffraction. Where there are major differences, the model must be adjusted or the electron density map refined. This process is repeated until the locations of all the atoms in the model agree with the computed electron density map. Max Perutz and John Kendrew developed this method in the 1950s; it was used to determine the structure of a protein for the first time in 1959 by Kendrew in his analysis of the protein myoglobin. Since then the structures of hundreds of proteins have been found in this manner.

All such structure determinations require computers, both to compute the electron density map and to manipulate the molecular model of the protein, but only the larger and more complex structures require a supercomputer. A case in point is the recent solution of the structure of an enzyme from a family called the kinases, which are critical in the regulation of cellular activities, by a team of researchers from the University of California at San Diego and the San Diego Supercomputer Center, led by Susan Taylor. Although the protein kinase family contains hundreds of members, until this work the three-dimensional structure of not a single member of the family had been known.

Calling on a Cray Y-MP supercomputer at the San Diego Supercomputer Center, the team used the molecular software X-PLOR, written by Axel Brünger of Yale University, to repeatedly adjust a model of the molecule until the computer-generated images of the protein structure agreed with the computed electron density map. Because of the molecule's complex three-dimensional shape, the use of the large stereo projection unit at the center's visualization laboratory was essential. Members of the team could examine the three-dimensional image as a group and share their insights into possible ways of improving the structure.

Fortunately for future researchers, the genetic sequence defining the catalytic core of the solved structure is common to all kinase proteins in eukaryotes; the structural agreement among kinase proteins again emphasizes the essential role evolution plays in modern molecular biology. The team's breakthrough will thus provide a "template" for the solution of structure for the other kinases. Once we have the three-dimensional structure of one protein from a family, it becomes possible to use a supercomputer to determine the structure of other pro-

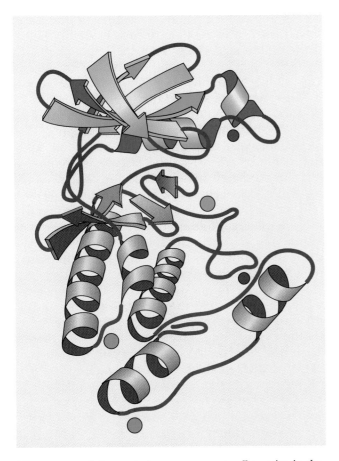

The structure of the catalytic core common to all proteins in the kinase family, as determined by x-ray crystallography aided by supercomputers and scientific visualization. The brown end of the protein can attach to the energy molecule ATP, while the purple end of the molecule binds to other amino acids. The green and purple dots indicate the location of insertions in the different members of this protein family.

teins in the same family through a technique called homology modeling.

An example of this technique in use can be found in the work of Bob Hermann and his colleagues at Eli Lilly and Company, one of the world's leading pharmaceutical companies. Hermann was interested in developing inhibitors for a family of enzymes called phospholipase A2 (PLA2). One member of the PLA2 family, human synovial PLA2, binds to cell walls and produces a variety of metabolic products, including arachadonic acid. This acid is then processed further, leading to molecules that are believed to be involved in human arthritis, rheumatism, and toxic shock syndrome. Hermann reasoned that if his team could determine the structure of human PLA2, then he could design a drug that would inhibit its metabolic action and prevent it from synthesizing arachadonic acid.

Although the sequence of amino acids in human PLA2 was known, its crystallographic structure was unavailable when Hermann started his work. However, as illustrated on page 119, the gene sequence that defines the human PLA2 protein has many bases in common with the gene sequence that defines the PLA2 proteins found in rattlesnake venom and in the pancreas of cows. The three-dimensional structures of both these animal proteins were known from x-ray crystallographic studies, so Hermann was able to use these structures as starting points. His group constructed a computer model of a first approximation to the PLA2 structure by taking the rattlesnake venom PLA2 structure, whose genetic sequence had 46 percent homology with the human PLA2 sequence, and replacing the 54 percent of nonmatching amino acids with the ones appropriate to the human PLA2 sequence, while keeping the protein backbone folded in the shape of the rattlesnake venom PLA2.

By placing this computer-generated molecular structure in a simulated water bath to recreate its biological environment and then running many hours of molecular dynamics on a Cray-2, first at the National Center for Supercomputing Applications and later on one purchased by Lilly, Hermann was able to relax the structure into one he was confident of. Knowledge of the structure enabled his group to move forward with their efforts to discover an inhibi-

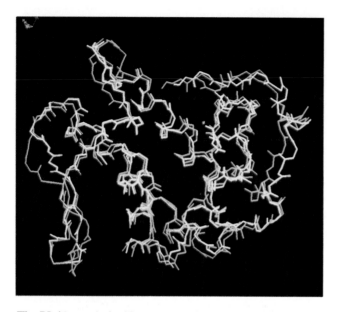

The PLA2 protein backbone computed by the homology method (orange) is superimposed over the protein backbone determined by x-ray crystallography (blue), showing a good though not perfect match.

tor two years before their laboratory colleagues obtained the structure of human PLA2 from x-ray crystallography. Indeed, Hermann helped speed the experimental effort by providing his model as the initial guess in the iterative procedure followed by the x-ray researchers, described in the kinase example above. The final proof of the method's validity is given in the image on this page, which displays the backbone of the predicted structure superimposed over the measured x-ray structure. Remarkably, this supercomputer-aided achievement was on the cover of *Nature* magazine in July 1991, the same month that the protein kinase structure appeared as a cover story in *Science* magazine.

Some proteins will not form crystals that can be examined by x-ray crystallography, and there may be no close relatives with known homologous structures. In this case, one is faced with the daunting task of computing the folded three-dimensional configuration of the protein from the linear sequence of amino acids. In theory, at least, it is possible to predict the folding of the amino acid chain from the interactions of the amino acids with each other and with the surrounding medium. Of all possible structures, the final structure will be the one with the lowest energy configuration. This "protein folding problem" remains the holy grail of computational molecular biology. However, it would take impossibly long for even the fastest supercomputer to model all the electrical forces that pull the linear sequence into a folded shape.

An easy way to see the difficulty is to compare the physical folding time with the speed of calculation. In a cell, the folding of a protein can take as long as a full second. Even running molecular dynamics codes for hours, a supercomputer can compute only tens of picoseconds of physical molecular motion, just a hundred-millionth of the time required for the protein to fold in a cell. The problem is that as the protein "searches" for the lowest-energy configuration during folding, making small adjustments to try one shape after another, it encounters many structures that are of lower energy than the previous shape, but which are not yet the structure of global minimum energy.

Still, it is clear from the above two examples that, by referring to already known structures, molecular dynamics can successfully adjust a guessed-at initial structure to a final form. However, when the percentage of bases shared with a known structure falls in the range of 25 to 40 percent, one enters what has been termed a "twilight zone." Peter Wolynes, of the University of Illinois at Urbana-Champaign, and his colleagues have developed a procedure that combines molecular dynamics with methods from information theory to solve the protein folding problem in this twilight zone region. Imitating the real protein, the molecular dynamics code wiggles the structure into new conformations, calculating the energy of each new form until it arrives at the structure of least energy.

The advantage of Wolynes's method is that it includes shortcuts to calculating the "energy function," as the relationship between form and energy is called. Using knowledge-based techniques similar to neural networks, described later in this chapter, his group has embedded the known structural motifs of proteins into an "associative memory" energy function. These motifs include common folding patterns that can exist along a piece of the amino acid chain, such as "alpha helices" or "beta sheets." Alternatively, the program can include information on whether a small stretch of amino acids is attracted to water molecules ("hydrophilic") or not ("hydrophobic"). The special energy function is designed so that as the molecular dynamics wiggles the structure, there is a deep energy minimum when some portion of the protein overlaps a previously known structural motif of a protein. Wolynes's procedure makes the descent to the configuration of minimum energy much more efficient.

Starting with the known linear sequence of amino acids, Wolynes, Zan Schulten, and Richard Goldstein could fold a protein in about one hour on NCSA's Cray-2. They have recently tested their method by folding proteins of known structure and then comparing the final folded forms with the experimentally determined structures. Their most impressive match was a protein, known as 2CRO in the Brookhaven Protein Data Bank, produced by a virus that attacks bacteria. Only 16 percent of this protein's amino acid sequence was identical with any of the proteins used in the associative-memory energy function. Yet when folded by their method, the protein takes on a final shape that is very close to its known structure. The question now is whether this method can be extended to larger proteins or to proteins whose sequences have even less overlap with those of known structure.

Protein Interactions

The static atomic structure of a protein that emerges from x-ray crystallography represents at best the average position of the atoms in a protein. The proper biological functioning of a protein requires that its atoms be in constant motion. Animal respiration presents an extreme case of the importance of atomic motion. Oxygen is carried from the lungs to the body's tissues by the protein complex hemoglobin. When the hemoglobin reaches the muscle tissue, the bound oxygen is released and absorped by a simpler molecule called myoglobin. The myoglobin exchanges carbon monoxide, a waste product of cells, for fresh oxygen.

Myoglobin has been well studied since it became the first protein to have its structure determined by x-ray crystallography over thirty years ago. The oxygen molecule binds to an iron atom, which is buried in the center of the globular protein to protect it from oxidation by external water molecules. Unfortunately, the only

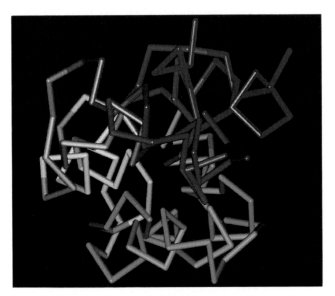

The shape of the viral protein 2CRO was derived using a knowledge-based method for protein folding. Placed over its structure is the structure of the same protein as derived from x-ray crystallography. Amino acids in the same position along the backbone are colored the same, from the first position (red) and to the last position (purple).

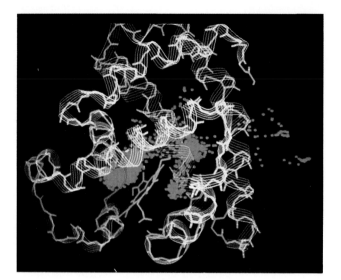

Carbon monoxide molecules (light blue dots) diffuse through a myoglobin protein (ribbon structure tracing out backbone), eventually emerging on the outside. The orange-red structure is the binding pocket where the diffusion starts. The picture sums up the positions of 20 carbon monoxide molecules over a period of 1000 picoseconds, revealing the internal cavities of the diffusion pathway.

way an oxygen molecule can cross the energy barrier represented by the intervening atoms and reach the central iron atom is by a rare quantum mechanical tunneling process. If the atoms of the protein were rigidly fixed in the positions given by the x-ray data, then it would take many times the age of the universe for one oxygen atom to tunnel through. But the oscillating atoms of myoglobin occasionally fall into a configuration representing a much lower energy barrier, and an oxygen atom is able to pass through. It is only because of the fluctuating motions of the myoglobin's atoms that we are alive!

Naturally, the appropriate technique to study these motions is molecular dynamics. Such studies inherently demand a supercomputer. Since the atoms in a myoglobin molecule move appreciably in as little as one femtosecond (one millionth of a nanosecond), the simulation

must resolve the atomic motions that take place over that length of time. Then it must compute 100,000 to a million of these femtosecond-long fluctuations to study the oxygen transfer. Early work by David Case from Scripps Clinic and Martin Karplus of Harvard University and their coworkers showed that the natural fluctuations of the myoglobin atoms would lower the energy barrier by about a factor of 10, allowing the oxygen to tunnel through in a time consistent with that obtained from experiment.

More recently, Ron Elber of the University of Illinois at Chicago and Karplus have followed the motions of an ensemble of 60 carbon monoxide (CO) molecules as these molecules escape from the inner cavity surrounding the iron atom in myoglobin. The pathways taken by the CO molecules tunneling out should be similar to the pathways taken by the oxygen molecules tunneling in. Elber and Karplus use a new computational technique that is a classical analog to the Hartree approximation used in quantum chemistry: each CO molecule feels the full force of the protein, but the protein feels the average force of all the CO molecules. As a result, one obtains 60 trajectories of CO molecules, each exploring an alternative diffusion pathway, while one only has to compute a single trajectory for the protein. This new method was incorporated into the widely used molecular dynamics code CHARMM, which was developed by Karplus and his coworkers.

Using the equivalent of 10 hours on a Cray Y-MP, Elber and Karplus simulated 100,000 steps of CO diffusion in the fluctuating myoglobin molecule. They found that the CO moves through an interconnected series of cavities in the myoglobin. When the CO enters a new cavity, it bounces around many times before it rapidly tunnels into another cavity or to the outside of the molecule. Thus, the motion of the ensemble of CO or oxygen molecules through a myoglobin protein is like a random hopping between a sequence of cavities. By the end of the 100,000 steps of simulation, half of the CO molecules had escaped to the outside of the molecule, while the others were still distributed along the various paths inside the protein.

Proteins interact not only with small molecules like oxygen and carbon dioxide, but also with larger macromolecules. One of the crucial functions of proteins is to act on the very DNA molecule that produced them, in order to regulate the production of other proteins. A major class of these regulatory proteins are the restriction enzymes. These molecules can recognize a stretch of the DNA nucleotide sequence and either bind to it or cut the molecule at an exact point. Performing this latter function they have become one of the biotechnology industry's most important molecular tools for cloning DNA.

For many years John Rosenberg's group at the University of Pittsburgh has been studying one of these restriction enzymes, Eco RI endonuclease. This enzyme protects bacterial cells by cutting the DNA of invading viruses. It cuts the virus DNA at fragments with the base sequence GAATTC. Rosenberg's group first refined the x-ray crystallographic structure of the protein using X-PLOR software on the Pittsburgh Supercomputing Center's Cray Y-MP. The group then turned to a molecular dynamics program called AMBER, and in collaboration with one of its authors, Peter Kollman of the University of California at San Francisco, they studied how the enzyme "kinks" the DNA at the right sequence before cutting it. They found that the protein wraps around the DNA, and that some of the protein folds are just the right shape to fit into the major groove of the DNA helix and spread its two strands apart. As we shall see next, the molecules of anticancer drugs can also interfere with the functioning of DNA.

The traditional search for a new drug begins with painstaking laboratory techniques to screen a variety of compounds collected from sources around the world. Promising drugs undergo years of tests in the test tube, on infected animals, and finally on human beings. Only one in tens of thousands of compounds originally screened is ultimately approved for human use.

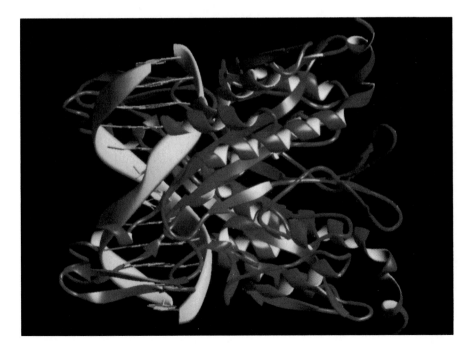

The restriction enzyme Eco RI endonuclease (red) wraps around the DNA helix (orange), then kinks the helix and cuts the DNA apart at that position.

Thus from start to finish, the cost of developing a single new drug may exceed $100 million. Supercomputers can help pharmacologists avoid many of the blind alleys that plague traditional drug development by focusing efforts on those drugs with promising structures.

Each year five million people in the world die from one of the large family of diseases termed cancers. Each of these diseases begins when a cell ignores the normal regulatory processes that restrain cell reproduction. Its uncontrolled replication leads to a clump of cells called a tumor. A fundamental problem in anticancer research is how to keep cells from dividing in a runaway fashion. Since cell division requires replication of DNA, many proposed strategies are aimed at blocking that process.

In a normal cell undergoing cell division, protein enzymes help "unzip" the double-stranded nuclear DNA. Other enzymes help replicate each strand to form a new copy of the DNA, then "rezip" the two full copies of the DNA, and move one copy each to the nuclei of the two daughter cells. Thus, cell division would be blocked if one could find a drug that either prevents the DNA from unzipping or stops the construction of new copies of DNA after unzipping. Anticancer drugs exist that accomplish one or the other of both functions. Supercomputers are helping scientists find related drugs that are even more effective or have less severe side effects.

Cyclophosphamide, one of the most widely used and effective anticancer and immunosuppressive agents, illustrates the first approach: preventing the unzipping of DNA. Frederick Hausheer and U. Chandra Singh of BioNumerik Pharmaceuticals in San Antonio, Texas, have used Cray Research supercomputers at the University of Texas, the National Cancer Institute, and the Scripps Clinic to model the behavior of the active portion of that drug, which is called phosphoramide mustard.

First, they determined the geometry of the chemical bonds in the drug in isolation by means of quantum chemistry computations. Sec-

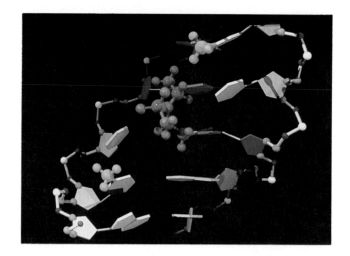

The drug phosphoramide mustard (in magenta) forms a crosslink that shackles together two strands of DNA, thereby preventing the cell from reproducing. The image is an average of the oscillating positions of the drug-DNA complex during 50 picoseconds, computed by molecular dynamics.

ond, they joined the model of the drug to a model of a DNA segment. Finally, they ran molecular dynamics programs that allowed the drug and DNA to adjust to each other's presence. As seen in the image on this page, the phosphoramide mustard forms a stable, strong chemical bond between the two strands of the DNA of a cancerous cell. This crosslink prevents the two DNA strands from unzipping, and when the malignant cell cannot divide, it dies. Similar work is also being pursued by researchers at Johns Hopkins and the Sandia National Laboratories.

Fighting Viral Infection

Viruses are parasites that must invade living cells to reproduce. Their fairly short genome of either RNA or DNA codes for a series of proteins that assemble into a protective coat surrounding the genetic material. The virus does not have its own machinery for transcribing the gene or building proteins. Instead, it takes over the existing cellular machinery of its host cell and puts it to use manufacturing new virus particles. A virus is thus a supramolecular structure composed of a complex of folded nucleic acid polymers and amino acid polymers, arranged in a compact and often geometric form.

Many of the structural details of viruses are revealed in the electron microscope. This instrument yields two-dimensional projections of the three-dimensional structure of the intact virus. Recently, the National Institutes of Health in Bethesda, Maryland, has created three-dimensional models with the help of a massively parallel supercomputer, the 128-node Intel iPSC/860. Calvin Johnson of the NIH's Division of Computer Research and Technology is collaborating with researchers Benes Trus and Alasdair Steven of the National Institute of Arthritis, Musculoskeletal, and Skin Diseases to study the geometry of the herpes virus.

First, the researchers digitize images of quick-frozen virus samples taken with a transmission electron microscope. The supercomputer determines the spatial orientation of each virus particle in an image, then combines the features from the images to produce a three-dimensional model of the virus. Although the three-dimensional structure can be computed on a minicomputer in about a month, the supercomputer performs the computations in only a few hours. As the speed of scalable parallel systems continues its rapid climb, researchers hope soon to achieve image resolution fine enough to reveal features of the viral surface that are crucial for designing antiviral drugs.

A very different approach to determining viral structure relies on the methods of x-ray crystallography. Using a Cyber 205 supercomputer, Michael Rossmann and his colleagues at Purdue University have created accurate electron density maps of a virus's protein coat from diffraction patterns generated by x-ray crystal-

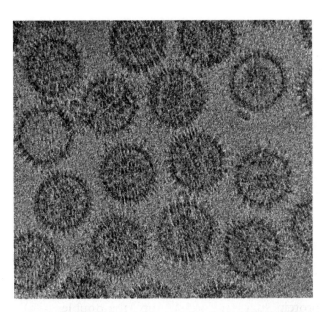

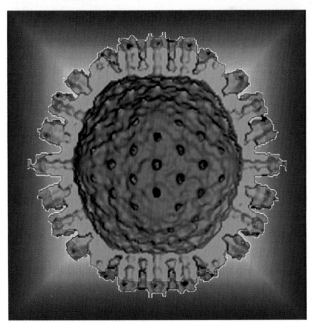

Top: *A digitized electron micrograph shows a number of antibody-labeled particles of herpes simplex type I virus embedded in ice.* Bottom: *The cut-away view of the computed three-dimensional image of this protein coat clearly reveals the regular shape of the surface protrusions and cavities. The red objects are not part of the virus, but are antibodies used as labels.*

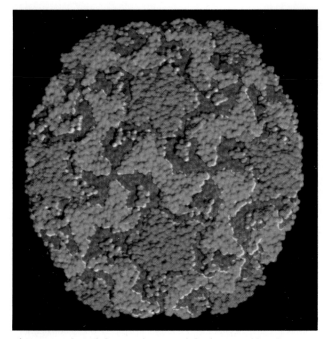

A computation of the protein coat of the human rhinovirus 14 at atomic level resolution, derived from x-ray crystallography experiments. The red represents the protein VP3, the green VP2, and the blue VP1. The virus contains about a million atoms and measures about 30 nanometers in diameter.

lography. They have investigated several members of the highly symmetric picornavirus family.

An image obtained in this way, reproduced at the top of this page, shows the human picorna rhinovirus 14 (HRV14), the cause of the common cold. The virus has an icosahedral symmetry; the protein coat is formed by 12 pentagonal caps, each composed of 5 equilateral triangles. Each triangle is built out of three proteins (called VP1, VP2, and VP3) locked together by a smaller protein (VP4) buried under the protein coat. With the virus structure in hand, scientists can begin searching the protein coat for features to be targeted by antiviral drugs.

One strategy for fighting viral infection is to produce antiviral compounds that inhibit viral replication. Careful examination by Rossmann of the HRV14 reveals a canyon 2.5 nanometers deep surrounding each of the virus's 12 vertices.

Rossmann and his colleagues conjectured that this canyon binds to specific receptor molecules on the surface of the host cell; this suggestion has since been confirmed by several research groups. A compound called WIN 51711, synthesized at the Sterling-Winthrop Research Institute in New York, is able to descend into these canyons and enter a small opening at the canyon's base. Once inside the opening, the compound lodges in a cavelike hollow. The compound gives the virus stability, inhibiting its ability to shed its protein coat and reproduce.

Using supercomputers, scientists are now systematically modifying the WIN molecular structure to minimize the binding "free energy" between the designed drug and canyon binding site. The American chemist Josiah Willard Gibbs, one of the founders of the scientific study of thermodynamics in the last century, showed that at constant temperature and pressure, all systems change in a direction such that the free energy is minimized.

A computational technique called the thermodynamic cycle-perturbation method allows one to compute the change in free energy as one replaces drug A binding to a target molecule with drug B. The computation typically involves transmuting several atoms of one element to those of another element and then using molecular dynamics to adjust the system. If the result is a lowering of the free energy relative to the original complex, then it is safe to assume that drug B binds more effectively to the target.

The usefulness of this method can be illustrated by considering the ability of the cold virus to evade antiviral drugs through mutation. Laboratory experiments have demonstrated that the mutation of just one amino acid in the coat protein VP1 can make the cold virus resistant to WIN drugs. Using NCSA Cray Research supercomputers, Rebecca Wade and J. Andrew McCammon of the University of Houston have shown that the binding of the WIN drug to the wild type virus has a lower free energy than the binding to the mutant virus, compatible with the experimental results. Wade and McCammon

The blue and yellow ribbon diagram represents the atomic structure, as determined by x-ray crystallography, of the protease enzyme encoded by the HIV-1 retrovirus. Bound to the enzyme is a peptide inhibitor (purple). This protease structure is the starting point for the supercomputing methods that determine which modifications of inhibitor molecules produce the most tightly binding inhibitors.

then alter the molecular structure of the WIN compounds on the computer and recompute the change in free energy; repeated passes could eventually produce a first guess at the most promising drug to combat the mutated virus. The great hope is that an arsenal of such supercomputing techniques will enable pharmaceutical companies to modify antiviral drugs as fast as the viruses are capable of mutating.

Similar techniques are used by Peter Kollman, David Ferguson, and Randall Radmer at the University of California at San Francisco as they seek drugs to fight AIDS. They are searching for compounds that will inhibit the action of certain protein enzymes created by the AIDS retrovirus. By computing the free energy changes in the enzyme-inhibitor complex, using the San Diego Supercomputer Center's Cray Y-MP, Kollman's group is attempting to determine which form of the inhibitor binds most tightly to the enzyme, thus best preventing replication of the AIDS virus.

Viral diseases may also be combated by developing antibodies that attack viruses. Antibodies are Y-shaped proteins that have trunks of similar molecular structure, while the upper arms and ends of the Y are formed of highly variable protein loops, enabling an organism to manufacture many different antibodies to bind

with a wide variety of target molecules or antigens. By developing a library of loops from the x-ray structures of many antibodies, scientists create a set of templates that can be manipulated by molecular dynamics and computer graphics techniques to model a large assortment of antibodies.

This technique is illustrated by the work of Shankar Subramaniam, then of Princeton University and now at NCSA and the University of Illinois, Scott Linthicum, Paul Kussie, and Jerry Anchin of Texas A & M University, and Jay Glasel of the University of Connecticut. They pursued a joint attack, coupling experiment with simulation, to discover an antibody surface that will bind the small molecule morphine. Using the Cray Y-MP supercomputer at NCSA and an array of Silicon Graphics workstations, they performed computations that clearly show how the antibody surface forms a concave cavity that is just the correct shape for docking with the convex morphine molecule.

There are exposed protein loops on upraised portions of a virus's hilly surface that similarly serve as antibody binding sites, as shown on page 130, but these loops are often altered as the virus mutates, rendering the existing antibodies impotent to bind to the virus. One clear strategy for attacking newly discovered virus

By searching for the antibody surface shape whose union with morphine produces the minimum free energy, the commercial code CHARMM is able to model the surface shape optimized for binding to morphine. Looking down onto the upper arm of one side of the Y in this image of the antibody, one sees each loop of the antibody protein in different colors, and the conserved structural portion of the protein in gray. The resulting cavity is just the right shape to bind to the morphine molecule, labeled M.

strains is to analyze how the mutations have altered the protein coat loop, then to design a slightly altered antibody that is again well matched to bind with the newly mutated virus. This is just on the frontier of what will be possible with the next generation of supercomputers. The computer graphics image of antibodies binding to a virus on the facing page illustrates the type of interaction that future simulations will be able to model.

Cells: The Building Blocks of Life

The cell is a remarkably complex entity composed of a large number of substructures and specialized organelles. Five billion protein molecules, selected from about 10,000 different types, account for roughly one-quarter of a cell by mass. The other three-quarters is mostly water and dissolved salts, like the original sea in which life evolved. The remaining few percent consists of small molecules such as lipids, amino acids, and the nucleic acids DNA and RNA. Even though the genetic material is a small portion of the mass of a cell, it is high in information content. To read out the genetic code, each cell has about 4 million ribosomes to manufacture proteins, and about 70 million tRNAs and 700,000 mRNAs present at any one time to carry genetic instructions to the ribosomes, where one million amino acids are bonded to growing protein chains every second.

Cells and their subcomponents have received comparatively little attention from supercomputing biologists in the past, but the subject is rapidly becoming an exciting frontier field. These biologists have focused most of their efforts on the study of important structural or metabolic complexes formed of interacting proteins and other molecules.

Of particular interest are the properties of the cell membrane, an extremely complex site of many processes critical for the support of life. The cell membrane comprises two layers of spe-

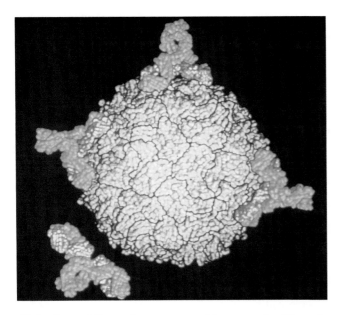

Molecular modeling and computer graphics expert Art Olson of the Scripps Research Institute created this image showing what antibodies binding to the outer envelope of the poliovirus might look like. Although the images of both the antibody and the virus are based upon experimental data and are computationally accurate, the interaction is the product of reasoned modeling. An accurate modeling of the process awaits the next generation of supercomputers.

cialized molecules called phospholipids: as can be seen in the image on page 116, each phospholipid has a long fatty tail pointing toward the center of the bilayer and a head forming the membrane surface. The 4-nanometer-thick bilayer provides a support for embedded proteins that create channels for the transfer of chemicals, anchor structural proteins, or catalyze critical chemical reactions.

Photosynthesis, the sunlight-powered reaction that transforms carbon dioxide into carbohydrates, is mediated by protein structures embedded in membranes. These proteins capture the energy that drives the reaction by converting an incoming photon of sunlight into stored electrical potential energy. Our understanding of this process took a great step forward when the three-dimensional atomic structure of the photo-

synthetic reaction center, a complex of four proteins and associated molecules such as the pigment chlorophyll, was deciphered for the purple bacterium *Rhodopseudomonas viridis* by Hartmut Michel, Johann Deisenhofer, and Robert Huber of the Max Planck Institute for Biochemistry in Martinsried, Germany. The reaction center contains the first integral membrane protein to have its x-ray crystallographic structure determined. This discovery in 1985 was recognized by the Nobel prize in Chemistry in 1988.

The photosynthetic reaction center acts as a biological photocell. Other membrane proteins capture sunlight and funnel the photons toward the reaction center. As a photon reaches the chlorophyll molecules in the heart of the reaction center, it excites an electron, which is conducted to one side of the membrane while the positively charged "hole" vacated by the electron migrates to the opposite side. The resulting charge separation produces an electrical potential across the membrane, thereby providing the organism with a source of energy for metabolism. Virtually every photon that strikes the photosynthetic reaction center successfully produces a charge separation. The liberated electron is always conducted in one direction through the membrane, since in traveling away from the reaction center, the electron loses energy and so cannot go back.

Molecular dynamics applications are one of the first to be run on new computer architectures, because of the simplicity of their coding. The National Institutes of Health's Center for Concurrent Biological Computing, in the Beckman Institute of the University of Illinois at Urbana-Champaign, directed by Klaus Schulten, has done a study of how a molecular dynamics code simulating the photoreaction center runs on a variety of architectures in their center and at the National Center for Supercomputing Applications.

For instance, it was found that the 12,000 atoms of the largest protein in the photosynthetic reaction center would not fit even in the large gigabyte memory of the NCSA Cray-2 using the standard CHARMM molecular dynamics program. By rewriting the molecular dynamics algorithm using a new method for summing up all the electrostatic interactions, Schulten's group was able to run the entire molecule on a smaller memory, massively parallel Thinking Machines Corporation's CM-2 supercomputer at NCSA, in much less time. Changing the algorithm can make an enormous difference in the research that can be carried out.

These simulations enabled Klaus Schulten and his colleagues to demonstrate that the electrical properties of the photosynthetic reaction center result from the overall structure of the macromolecule, rather than from certain side groups of atoms, as had been hypothesized earlier. This discovery explains the extraordinary fact that genetic mutations, even those that significantly alter the arrangement of various structures within the macromolecule, do not adversely affect its basic ability to perform photosynthesis.

Schulten, in collaboration with Helmut Heller and Michael Schaefer from his center, has turned his attention to a smaller photosynthetic protein, bacteriorhodopsin, which developed earlier in the evolutionary history of bacteria. Using an array of 60 parallel transputers, specialized computing intensive boards that can be attached to a workstation, the group was able to compute a molecular dynamics simulation of the entire membrane-embedded protein, including a thick layer of water molecules lying on either side of the cell membrane. This complex supramolecular structure, shown in a visualization produced by Andreas Windemuth of the same group, is shown on page 116. Such computations point the way toward more realistic simulations that would capture a variety of tightly interacting elements in the cell.

The fluid inside a living cell is distinguished from the fluid outside by large differences in the concentration of certain key ions across the cell membrane, such as sodium, potassium, calcium, and chlorine. For example, the concentration of the sodium ion is 20 to 40 times lower inside a cell than in the bloodstream, and the concentration of the calcium ion is 1000 times lower. This difference in electrically charged ions creates a large potential difference across the cell membrane, which can be as high as 100,000 volts per centimeter! To maintain this charge separation takes up to 50 percent of the energy stored in ATP molecules, the energy reservoir of cells. The regulation of this ionic balance has recently begun to be studied using supercomputers.

The ions move across the cell membrane through narrow ion channels made of protein. One of the simplest ion channel proteins is gramicidin A, a protein that consists of a long helix that penetrates the cell's lipid bilayer membrane; like a pipe, it connects the inside and the outside of the cell. In 1989, See-Wing Chiu and Eric Jakobsson of the University of Illinois at Urbana-Champaign, in collaboration with Shankar Subramaniam and J. Andrew McCammon of the University of Houston, used the NCSA Cray Y-MP supercomputer to study the molecular dynamics of water molecules moving through this helix. They found that the water molecules could move through the channel only one at a time. The chain of water molecules interacts with itself and the atoms of the channel wall to produce the correct rate of ion and water transport.

Most simulation studies to date have focused on the details of a specific function such as the ion channel flows, but recently programs are being developed that model the unified interaction of various subcomponents of the cell. Unified models of ion transport sum all ion transport for an entire cell: using the results of "side calculations" of the details of the channel transport, together with the chemical reaction rates, these models attempt to understand how the cell can maintain its huge ionic balancing act.

As an example of this approach, Janet Novotny and Eric Jakobsson, working at the National Center for Supercomputing Applications, have recently constructed a model of the thin layer of cells, or epithelium, lining a lung airway. This model elucidates how the symp-

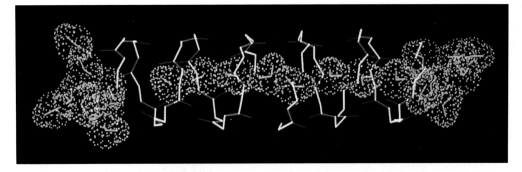

A snapshot of water molecules moving through a gramicidin helix ion channel shows their preferred orientations. The spheres formed of blue dots show the size of the atoms in the water molecules.

toms of the disease are produced by the cystic fibrosis genetic defect, which disables membrane proteins that transport chloride ions into and out of the cell. Their simulation shows how the disabling of these membrane proteins renders the epithelium incapable of water transport, which leads to the clogging of the lungs by thick mucus.

The Circulatory System: Computing the Flow of Blood

Aggregations of cells form the body's tissues, and tissues of many kinds cooperate in the functioning of the interconnected systems that compose living organisms. The circulatory system provides a good example of how supercomputing is beginning to be useful for modeling the components and activity of these systems.

The circulation of blood supplies oxygen and nutrients to every living cell in the body, removes metabolic waste products, and transports heat to and from cells. An elaborate and versatile computer model of the human cardiovascular system was created in the late 1980s by Marlyn E. Clark at the University of Illinois at Urbana-Champaign and his colleagues Jan Duros and Robert Kufahl. Their model consisted of a network of 96 arteries assembled in a whole-body configuration. The arteries were divided into more than 600 segments for the purpose of computation. A judiciously chosen parameter quantifying a resistance was added at the ends of the vessels to simulate the effects of all the vessels not modeled.

Two sets of equations describe blood flow, one expressing the conservation of mass and the other the conservation of momentum. The so-called continuity equation expressing mass conservation says that the difference between the inflow and outflow from a small segment of elastic vessel equals the fluid stored in that vessel. The continuity equation thus tracks the volume of blood in the system's 600 segments. In cases like blood flow where the force exerted depends not only on variations in pressure but also on viscosity, momentum conservation is described by the Navier-Stokes equation. Both equations were incorporated in a finite-differenced scheme that calculates pressures and rates of flow throughout the vascular network. The resulting computer algorithm can calculate the effects of tapering, branching, and multiple looping in the arteries as well as the effects of aneurysms and vessel constrictions. Included in this model is a parameter, called a compliance coefficient, that expresses the elasticity of the arterial walls. This parameter can be adjusted to simulate walls that range from flexible and elastic to very stiff and arteriosclerotic.

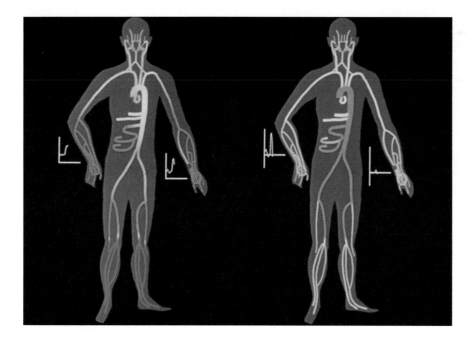

This image, one of 120 evenly spaced throughout a heart beat, displays values of blood pressure (left) and flow (right). High blood pressure and flow are indicated by red hues and low blood pressure and flow by blue hues. Intermediate values of pressure and flow are indicated by intermediate colors on a rainbow scale from red to blue. The graphs track alterations in pressure and flow with time at the wrist.

Using a Cray-2 supercomputer, Clark and his colleagues calculated blood pressure at 753 nodes and the rate and volume of flow at 685 nodes throughout the 96 arteries of their whole-body model. To display these data, Michael McNeill at the National Center for Supercomputing Applications used Wavefront software to produce a video of animated sequences showing changes in pressure and flow as the blood courses through the arterial system.

When a pulse of blood reaches a vascular junction or the bifurcation of an artery, the blood is reflected back when it hits the channel wall or a branch point, producing waves that move backward through the circulating blood. Clark's simulation demonstrated that highly elastic blood vessels readily produce copious reflected waves that persist over long distances and make complex interference patterns. Stiffer vessels produce far fewer waves, but they travel faster. Since factors like wall stiffness are not easily altered in laboratory experiments, supercomputer simulations are becoming an increasingly important tool in cardiovascular research.

Clark's simulation also probed the factors leading to aneurysms in the brain. Sometimes the wall of an artery gives way under stress, and an arterial segment expands to form a balloon-like dilation, the aneurysm. A sudden surge in blood pressure can rupture the aneurysm, sending masses of blood into the brain and causing severe neurological damage or death. Most commonly, intracranial aneurysms are found in the circle of Willis, an arterial interchange through which virtually all blood that nourishes the brain must pass. Clark's simulation reveals details of events that can lead up to an aneurysm's rupture. In order to create a huge short-duration spike in blood pressure in the circle of Willis, several pathological conditions must exist: hypertension, stiff (arteriosclerotic) arteries, and numerous reflecting sites both near and far from the brain.

Blood is kept in motion by the beating of the human heart, a large, hollow muscle that pumps about five quarts of blood through its chambers every minute. Each side of the heart performs a different pumping function: the right

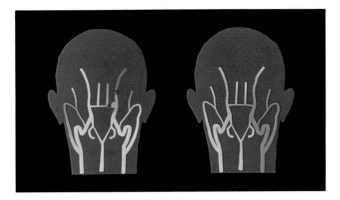

At the left, a snapshot shows the circulation pressure near two cerebral aneurysms at a particular moment in the cardiac cycles. The red dots mark the positions of the aneurysms, which reduce the blood pressure to various portions of the brain. On the right is a comparative view of normal blood circulation.

side takes blood from the body and pumps it to the lungs, and the left side collects blood from the lungs and pumps it to the rest of the body. The two sides of the heart relax and fill at the same time during the diastole phase; they contract and empty themselves during the systole phase. The action felt as a heartbeat is the systole. Supercomputers are being used in a variety of ways to study this organ: they play a role in observing its dynamics, in modeling its blood circulation, and in designing substitutes for it.

To aid in the diagnosis and treatment of heart disease, physicians would like to have high-resolution images showing the heart in action along with relevant quantitative information, such as the volumes of the ventricular chambers and the thicknesses of the walls. In effect, physicians need a four-dimensional x-ray machine. As futuristic as such a device sounds, it is close to existing in practice.

Computer-aided tomography (CAT) is a technique that uses a pencil-thin beam of x-rays to produce a cross-sectional view through objects like the human body. The body is placed between the x-ray source and a detector, and the two devices are slowly rotated about the body in such a way that a particular slice of the body is illuminated from all angles. The detector measures the intensity of x-rays penetrating the body, and a computer combines data from all angles to produce the cross-sectional image. Many two-dimensional slices can be combined to make an image in three spatial dimensions. Only a few specialized CAT scanners can make a series of three-dimensional images at closely spaced intervals of time. One of these is the Dynamic Spatial Reconstructor (DSR) at the Mayo Clinic.

Eric Hoffman, at the University of Pennsylvania, used this device to produce a CAT scan of the beating heart inside a dog. The DSR can generate 60 image cubes per second, each of which has two million volume elements. Hoffman, in collaboration with Thomas Huang of the Beckman Institute, brought the data to the NCSA biomedical imaging group headed by Clint Potter. Using the Viewit software they had developed, the group created images with a Thinking Machines Corporation CM-2 supercomputer and then viewed them on a Silicon Graphics workstation. An interactive tool called Tiller, developed by Pat Moran of NCSA, allows the user to run the images as a digital movie, whose orientation can be changed at will while the image of the heart is beating.

It will take some years yet before scanners like the DSR are linked with adequate computer power so that physicians can view details of the functioning heart instantly. The scanner, the supercomputers, and the software are all research prototypes. Still, we can begin to imagine the future impact of such combinations of supercomputers, networks, and scientific instruments on both research and clinical practice. Ultimately, our whole bodies may be scanned and recorded digitally during our annual checkups. Artificial intelligence programs could compare our organs and physiology with the many others in a vast national data bank to detect immediately significant deviations from our past condition or from national averages. The realization of this vision lies in the next century, but already some significant progress is being made.

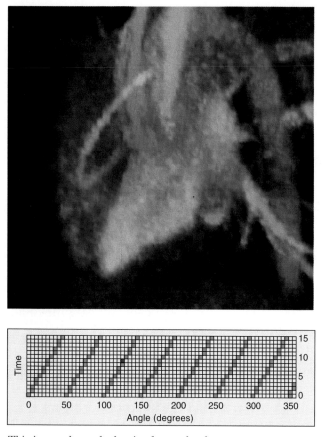

This image shows the beating heart of a dog as seen on a Silicon Graphics workstation running Tiller. The image is from a movie of three-dimensional CAT scans that have been visualized at each moment in time (vertical axis) from all viewing angles (every five degrees-horizontal axis). The user specifies with a mouse the sequence of angles to view in time (blue squares), then the sequence of images runs as an animation on the screen.

Another approach to visualizing the heart's activities has been successfully carried out by Charles S. Peskin and David M. McQueen at New York University, using supercomputers to model blood flow in the human heart. In setting up their work, Peskin and McQueen have had to overcome the challenge of simultaneously modeling components with very different properties. On the one hand, heart muscle is an elastic material with properties that change over the

heart-beat cycle; heart muscle is considerably stiffer in systole than in diastole, for instance. On the other hand, the blood in intimate contact with the heart muscle is an incompressible, viscous fluid. Therefore, in modeling the behavior of the fibrous walls and valves of the heart along with the concomitant fluid flow, Peskin and McQueen tackled a problem in both the hydrodynamics of fluid flow and the elasticity of a solid material. They developed a computer code that computes the forces generated by a system of muscle fibers under tension, then coupled the results with the Navier-Stokes equation to give a description of the blood flow. These equations were solved with finite difference methods on a lattice 64 zones on a side, using both a Cray Y-MP at the Pittsburgh Supercomputing Center and a Cray-2 at the Minnesota Supercomputer Institute. Representative results showing computed flow patterns are seen in the images on the facing page.

A simulated heart can be used to test artificial valves. Peskin and McQueen have been able to improve the design of a single-pivoting disk valve and have received a patent for their design of a butterfly bileaflet valve. The simulated heart can also be used to study physiological issues such as the effect of varying the delay time between atrial contraction and ventricular contraction. Certain pacemakers pace the atria and the ventricles separately, and these studies give an optimal delay time to program into such devices. Finally, fluid dynamic simulations have been used by other groups to study completely mechanical hearts. The results point to flaws in the design that may create turbulent flow damaging to blood cells or stagnate flow conducive to clotting.

The Human Brain

The final frontier of biology is undoubtedly the human brain. It shares the physical characteristics of other systems of the body, in that it is

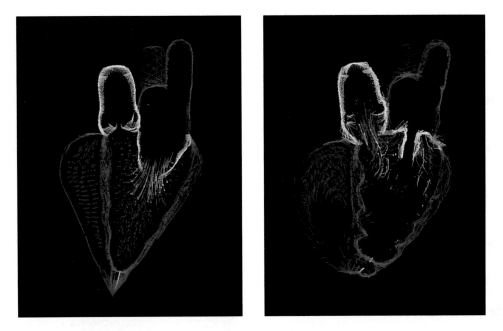

At the beginning of diastole, blood flows into both atria; the tricuspid and mitral valves are closed. The atria contract, forcing the blood through the mitral and tricuspid valves into the ventricles (left). The filled ventricles then contract and the resulting pressure shuts the mitral and tricuspid valves while opening the semilunar valves, allowing the blood to flow into the aorta and pulmonary arteries (right). As the heart relaxes, the semilunar valves close, the atria expand and fill with blood, and the cycle is repeated.

made up of proteins, cells, and blood networks, but it also has a profound information processing capacity of its own. The brain accepts information from sensory receptors, restructures it to create perception, manipulates it in the process of thought, and acts on it through the control of muscles and glands. The immensely varied activity of the brain is made possible because developing neurons take on different functions depending on the different connections they have established to other neurons.

Neurobiologists seeking to understand the brain are faced with the task of deciphering the functions of over a trillion neurons, each of which may form as many as a thousand connections with other nerve cells. Some experts estimate that even the teraflop supercomputer will be a thousand times slower than a human brain,

and organized in a much less interesting fashion. So daunting is the brain's complexity that for decades to come we can expect to achieve only partial insight into the mysteries of its functioning. Supercomputers are being used to help laboratory microscopists create images of the static cellular neural tissue, as well as to build models in the form of neural networks of how large numbers of neurons interact with each other.

Magnetic resonance imaging (MRI) is a particularly effective technique for imaging the brain because it can show soft tissue at a variety of scales. MRI makes use of the fact that protons in the nuclei of atoms possess a tiny magnetic field. If a substance is placed between the poles of a strong magnet, the external magnetic field causes the protons to precess, or spin around, at a specific frequency. If the substance is also exposed to radio waves at that same frequency, the precessing protons absorb energy from the radio waves. A reconstructed image of the absorption frequencies then maps the density of protons in the sample, capturing the varying densities of the soft tissues in the body.

Paul Lauterbur, one of the inventors of MRI, has recently been collaborating with his colleagues Xiaohong Zhou and Clint Potter, of the University of Illinois and the National Center for Supercomputing Applications, to use MRI and the NCSA supercomputer network to image details of the brain smaller than ever before captured with that imaging technique. By using a special research MRI unit, he can image microscopic areas of brain tissue in three dimensions. The MRI sends the measured absorption frequencies to the Cray-2 supercomputer, which constructs images in a few minutes.

The three-dimensional transparent image is then sent back to a graphics workstation, where it can be observed from all sides as it rotates around its vertical axis. The microscopic MRI image on this page shows a sample of brain tissue less than 1 millimeter in diameter. Observation of the same sample through an optical microscope verified that the solid objects appearing in the MRI image are indeed the nerve fiber tracts and the capillaries in the tissue sample.

In a similar fashion, neurobiologists Mark Ellisman and Steve Young at the University of California at San Diego have used a supercomputer to help reconstruct images of individual nerve cells from the brains of patients suffering from Alzheimer's disease. Data acquired with an intermediate high-voltage transmission electron microscope at the San Diego Microscopy and Imaging Resource provided two-dimensional images of a series of brain tissue slices 0.25 to 5 micrometers thick. Software written by David Hessler was used on the Cray Y-MP at the San Diego Supercomputer Center to convert the slices into the three-dimensional image seen on the facing page. Under development is a software program, termed the Microscopist's Work-

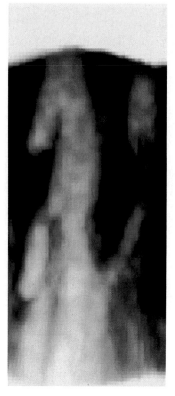

This image of nerve tissue in a cylindrical glass tube 800 micrometers in diameter was constructed by coupling a Cray-2 over a high-speed network to an MRI unit in Paul Lauterbur's Biomedical Magnetic Resonance Laboratory. The image is 64 zones on a side, and each zone is 13 micrometers across.

station, that allows remote users to transfer, browse, study, and analyze many types of microscopy images from their desktop computers.

One can imagine these early efforts leading eventually to a "digital microscope": the user will move the image around with a mouse at a desktop computer, while the sample is continuously observed at a remote instrument. Prototypes are being set up at the Beckman Institute, where computers are connected to three different scientific instruments (a scanning tunneling microscope, a magnetic resonance imaging device, and a laser scanning confocal microscope), and at the San Diego Microscopy and Imaging Resource in collaboration with both the San Diego Supercomputing Center and the Scripps Research Institute.

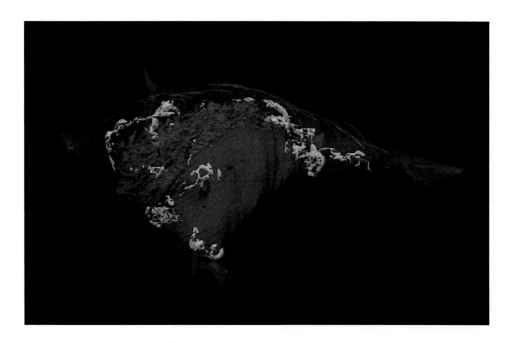

An image of a single Alzheimer neuron. The three-dimensional surfaces are reconstructed from the intersection of the surfaces seen on nearly 100 separate two-dimensional slices observed with the electron microscope. Paired helical filaments (orange), characteristics of the disease, appear to obstruct interaction between the nucleus (purple) and the Golgi apparatus (white), where sugars are added to newly synthesized proteins.

In 1949, Donald Hebb, a psychologist at McGill University in Montreal, argued that for the brain to learn a new concept or task, some physiological change must occur in its tissues. Specifically, he suggested that the passing of a signal between two neurons strengthens the synaptic connection between them. This brilliant conjecture, now supported by decades of neurobiological research, is the underlying idea behind a theory of learning that can be simulated using algorithms called neural networks.

A large-scale neural network is a dense web of interconnected processors. At first the connections between the processors can be quite random. But as input is processed by the computer, the strength of connections between specific pairs of processors changes according to Hebb's rule that links grow stronger with use. Information in a neural network is stored not like bits of data in a memory chip, but as the strengths and patterns of interconnections between processors. Rather than carry out a fixed computer program, as the computers in most

examples in this book do, a neural network "learns" to perform tasks by being shown a series of examples. As a result, neural networks have the uncanny ability to spot patterns in mounds of data. They can learn to understand speech, recognize objects, and make shrewd guesses based on incomplete data. These traits suggest that neural networks may be an important path to the achievement of artificial intelligence.

In animals, sensory input—be it from the eye, the ear, or the skin—is often mapped onto a thin layer of cells in the brain. These layers contain two-dimensional patterns called brain maps, created as groups of neurons become electrically active in response to properties of the sensory input. Thus, brain maps contain basic information about what the animal is seeing, hearing, or feeling. These maps have actually been observed in the visual cortex of the macaque monkey. A small hole created in the back of the monkey's skull exposes a patch of the visual cortex; to this patch are applied

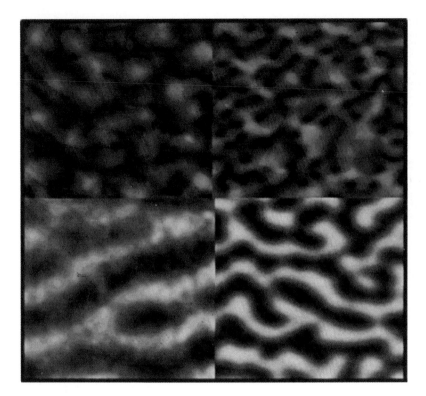

Left: *Observations by Gary Blasdel of a small portion of the visual cortex of a monkey. The upper half shows the sensitivity of cortical neurons to orientation, with red representing vertical lines and green representing horizontal lines. The bottom image shows sensitivity to visual stimulation of the right (light) or left (dark) eye. One sees that the brain spatially interlaces images from both eyes.* Right: *The same stimuli and color maps are shown here from the supercomputer neural network. The similarity of the spatial patterns indicates that the neural network has captured the essential aspects of the visual cortex organization.*

voltage-sensitive dyes. These dyes change their spectral colors as a function of the local electric potentials and permit researchers, such as Gary Blasdel of the Harvard Medical School, to photograph electrical activity on the visual cortex as the animal is shown lines oriented at various angles. The resulting images of the visual cortex, such as those shown on this page, can relate the spatial organization of the neurons in the visual cortex with the properties of the visual stimuli.

To understand how information is actually being represented in these brain maps, it is first necessary to know how the brain processes sensory input as it passes from the retina to the visual cortex. This processing is not genetically determined in its entirety; the pattern of strengths and weaknesses in neuron connections emerges as a result of an organism's interactions with the environment over the lifetime of an

individual. To try to capture this process of self-organization, Klaus Schulten and his colleagues Klaus Obermayer and Helge Ritter in the Beckman Institute at the University of Illinois constructed a neural network algorithm that builds a connection scheme between the processors in the Thinking Machines Corporation's CM-2 supercomputer in a way that simulates synaptic connections in the brain. The algorithm will use this connection scheme to correlate positions on an input map, called the receptor surface, to positions on an output map, called the cortical sheet.

Initially, thousands of random connections link the processors in the CM-2, and so the output map is featureless, regardless of whatever pattern is on the input map. The neural network is then trained by placing patterns on the input map while the connections between the processors in the Connection Machine are modified according to Hebb-like rules. Interconnections that are used frequently are strengthened, while those used infrequently are atrophied. After about 60,000 adaptive steps, requiring many hours on the CM-2, the output map is strikingly similar to that actually observed in experiments. In this way, Schulten and his colleagues have discovered a model for how a specific portion of a macaque monkey's brain might be wired.

Laboratory experiments with cats yield slightly different patterns on the visual cortex.

To simulate these patterns, Schulten and his colleagues must adjust the parameters of the Hebb-like rules that modify the connections in the CM-2. When the output map finally resembles patterns on the feline visual cortex, the connection scheme in the supercomputer represents a likely model of the wiring of the cat's brain. Furthermore, by comparing the connection schemes in the CM-2 for the monkey and the cat, scientists can study basic differences in the electrical circuitry in the brains of these two different animals.

Engineering Design and Analysis

Before the advent of supercomputers, industrial designers relied solely on models and prototypes fitted with instruments to measure quantities such as stress and strain. In contrast, today's engineer sitting at a graphics screen connected to a supercomputer or workstation can directly observe the effects of the forces acting on the simulated components under design. The engineer can modify a design component by clicking a mouse and promptly see the changed distribution of stress and strain as well as other variables.

As the engineer shaves material from one section or adds a support strut to another, the finite element mesh is adjusted appropriately, and the program recomputes the stresses and strains, telling the engineer whether the alterations were for better or worse. By working closely with a computer, the engineer is able to maximize the strength, safety, and durability of components while minimizing the

Simulated rays of light emerge from openings in the cab of an earth-moving machine, defining the driver's field of view. The periodic light stripes help clarify the emerging light bundles.

cost, time, and materials needed to manufacture them.

Many of the engineering problems that we will study in this chapter were solved using commercial codes that perform finite element analysis. These codes run on a variety of machines from personal computers to supercomputers; many of the leading codes run in parallel on vector multiprocessors, particularly those from Cray Research, Convex, and IBM. Codes for massively parallel machines are still being developed, but commercial versions may become available shortly after this book is published. Advanced features of the commercial codes include automatic mesh generation and refinement, interactive graphics, sensitivity analysis, design optimization, and more advanced physics such as interactions between gas flows and elastic deformations or fluid-solid body interactions. Whereas codes running workstations or the more powerful personal computers can perform computations on tens of thousands of finite elements, the most detailed supercomputer analyses can reach hundreds of thousands of elements.

Supercomputers are often thought to be useful only to large companies such as automobile and aerospace manufacturers or to government laboratories constructing enormous projects such as fusion reactors. Although such applications are quite important, the real story in computational engineering is its use to improve design of a broad range of consumer and industrial products. We start the chapter with three recent examples of such efforts: designing better golf clubs, diapers, and bulldozers.

Improving Your Swing

When the MacGregor Golf Company decided to bring out its first golf club made of the exotic metal titanium, it teamed up with Cray Research Corporation and one of the major engineering software houses, MacNeal-Schwendler, to study how to produce the best design. New

software modules had to be written to create finite element models for the ball and club separately, to model the forces on impact, and to handle the deformations and oscillations of the golf ball in flight, characteristics that alter the ball's aerodynamics and range. Oddly enough, the golf ball required ten times the number of finite element nodes as the club head did because of its dimples and the relatively larger deformations it experiences.

After twenty hours of computation on a Cray Y-MP, the simulation shows in detail how stresses develop on the surface of and inside the club as it strikes the ball. The analysis led to changes in the design of the club head; for example, slots and gear teeth were cut into the titanium to stiffen the club, reduce stress, and improve the aerodynamics. The new design gives the average golfer a bigger "sweet spot" on the club head and hopefully a straighter tee shot.

The supercomputer simulation offered a number of advantages over the normal testing procedure. Designers cannot gain much information by videotaping the golf club hitting the ball, since for a typical swing that impacts at 100 mph, the ball moves five feet while one 1/30-second video frame is shot. A strain gauge attached to the club measures strain at only one point, whereas the supercomputer simulation captures stress at 1000 locations in the head at each moment in time. Furthermore, the analysis took only three weeks, instead of the typical three months required to test a prototype conventionally. The results of the analysis were unveiled not at a scientific conference, but more appropriately at the 1991 U.S. Open.

A Drier Diaper

Developers of consumer products have begun investigating the use of supercomputer simulations only recently. Their new-found appreciation of the technique is perhaps best illustrated

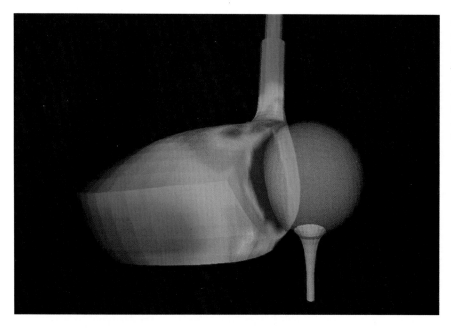

A view of a golf club hitting a ball shows how stress is distributed at the moment of impact. Colors indicate stress levels from red for highest stress to purple for lowest. The stresses were computed using the commercial code MSC/DYNA, and the results were transformed into images using the Multipurpose Graphic System developed by Cray Research.

by the effort to design a more absorbent disposable diaper at the Dow Chemical Company. Plans were underway to incorporate new superabsorbent polymer particles called Drytech™, developed at Dow. In traditional laboratory tests, liquid was poured on the surface of the Drytech-filled diaper and the extent of the wetness was measured.

Dow researchers realized that they could create a mathematical model of the diaper, which would include the outer covering, normal fluff fill, and a distribution of Drytech particles. Their simulation followed the stages of liquid movement through the diaper. First, the liquid is quickly transported through the diaper by capillary action; consequently, the thickness of the pad is reduced by the surface tension forces of the now wet diaper. The distributed liquid is then taken up more slowly from the cellulose fluff by the superabsorbent particles. Eventually, the liquid moves through the pores between the fluff and the fluff re-expands.

The visualization of the simulation shown on the next page, which was performed on a Cray-2 supercomputer at the National Center for Supercomputing Applications, allows one to see clearly the spatial extent of the liquid and the distribution of the particles that did most of the absorbing. The computation gave results that were in good agreement with experiment, but now the researchers had a "what-if" computing program that allowed them to rapidly try out a variety of different designs. The laboratory tests could then provide the final proof of design. This procedure substantially shortened the development process and produced a more reliable diaper.

Designing a Better Bulldozer

Human factors are critical in the design of large machines piloted by people. The machinery for moving heavy earth, for example, endures constant pitching and turning as it rides over very rough terrain. Throughout this time, the human operator must have the widest possible field of view to ensure the proper use of the equipment.

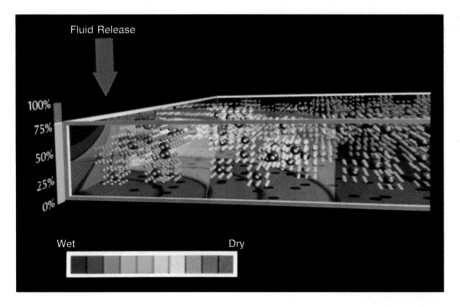

Fluid Release

100%
75%
50%
25%
0%

Wet   Dry

This cross-sectional view through a diaper shows the absorption of fluid 12 seconds after release. The color on the base and on the particles indicates the relative amount of liquid present. The grey balls are the particles of Drytech™, their sizes showing how much liquid has been absorbed; the other particles are the fluff filler.

Thus, a design conflict is set up: the operator has the best view when there is no vehicle around him, but the vehicle structures are required to perform the tasks at hand.

Researchers at Caterpillar, Inc., worked with Mark Bajuk and other graphics professionals at the National Center for Supercomputing Applications, to develop a number of innovative visualization techniques that allow designers to optimize visibility. A database from Caterpillar contained the specification of the vehicle's proposed design. Using Wavefront software on a cluster of Silicon Graphics workstations, the team was able to create an animation that showed what the operator could see as the machine carried out its chores of lifting and dumping. When the team detected a vehicle configuration that seriously blocked vision, they modified the database defining the vehicle and had the visualization recomputed. This visual "debugging" of a design required massive computations to create the animation, yet was much faster than the old method of actually building the new vehicle models in order to evaluate the design.

A novel approach to mapping the visual space of the operator is shown in the image on page 144. A simulated light bulb is placed in the simulated operator compartment at the eye level of the operator. The software then traces the outgoing cones of light framed by each window in the compartment. Any obstacles cast shadows, shown in dark grey. The resulting image clearly reveals in one picture the "blind spots" created by the current design. It is then straightforward to modify the design database and recompute the consequent changes in the visual field.

Virtual reality devices at NCSA, which compute a stereo three-dimensional view in real-time and project it into the viewer's eyes, now allow designers to actually sit on a model driver's seat and observe the view allowed by the current design. Controls have been added so that the designer can literally operate a virtual earth-moving machine. As these techniques are adopted by other manufacturing companies, we can expect safer human-operated machine products to emerge.

Process Engineering: Casting Steel

Many companies manufacture machinery that is used by other industries to create their products.

Two good examples are steel objects shaped to serve as components in more complex products and machines that sterilize cans of food. In both these cases, engineers focus on delivering the product within tolerances, with no defects, and with as little waste as possible. Where it may be impossible to put probes inside the material being processed, supercomputing and visualization can reveal design flaws hidden from view.

Although casting dates back to the Bronze Age, it is the seventh largest industry in the United States, because it provides basic parts for many other industries. Engineers wishing to avoid defective products are highly motivated to understand precisely how molten metal solidifies in a mold. A simulation by Jonathan Dantzig, a professor at the University of Illinois at Urbana-Champaign, studied heat transfer and the process of solidification during the foundry casting of a steel compacting hammer. Measuring about 18 inches on a side, the compacting hammer is part of a machine that crushes automobiles. At a foundry, the hammer is cast by pouring molten steel into a sand mold, whereupon it cools and solidifies in a complex fashion.

The simulation was performed on a Cray X-MP at Cray Research using the commercial finite element code FIDAP from Fluid Dynamics International. For his simulation, Dantzig filled the inside of the mold with a complex mesh of 7000 finite elements in the form of hexahedrons, tetrahedrons, and wedges. The surface was covered with quadrilaterals. Both the hammer and a structure called a riser, which protrudes from the top of the hammer, were modeled. The riser serves both as a portal through which steel is poured into the mold and as a reservoir of molten metal. Steel contracts as it solidifies, and so molten metal in the riser is gradually drawn down into the mold as the hammer cools.

In the two stills from an animation of the simulation on this page, a contracting grid inside the hammer shows the location of the boundary between solid and partially molten steel. In the second, later view, we see that a pool of molten steel has separated from the reservoir in the riser. When this isolated pool of steel solidifies, it will leave a region of lower density inside the hammer. Dantzig's simulation has therefore

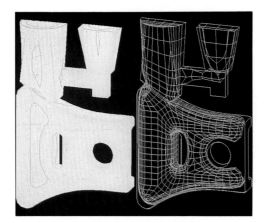 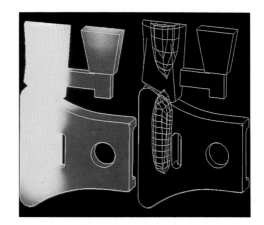

Color displays the surface temperature of the cooling steel in these two images from a simulation of a foundry casting, which show cooling from white hot to red hot. Twelve and a half minutes after the steel was poured into the mold (right), the molten steel in the hammer has separated from the reservoir at the top. The changing temperature on the surface of the steel (yellow is hotter, red colder) gives no clue that this potential defect is occurring in the casting.

revealed a significant flaw in the design of the mold that would result in an unacceptable product.

In this example, the molten steel is relatively immobile as it solidifies, as is also true for the casting of steel sheet, pipes, beams, and plates. In contrast, hot-rolling processes require hot ductile steel to move rapidly down a production line while being squeezed by electronically controlled rollers into the correct thickness and width. A group led by Jerry Kahrs, of the USS Technical Center, the research and development arm of USX's steel-making group, has been working with the Pittsburgh Supercomputing Center to develop ways to use high-performance supercomputing not only to create models, but ultimately to control the manufacturing process itself.

Kahrs was able to handle the complexities of the problem using the finite element code DYNA3D, developed at Lawrence Livermore National Laboratory in California. The code tracks the temperature and chemical composition of the steel, the deformation of the milling machinery, the heat transfer by both conduction and convection, and the smoothness of the sliding of the hot steel between the rollers. By analyzing all these factors, it is able to accurately predict the state of the rolling steel. Because the effects of temperature change on the behavior of the steel can be quite strong, the flow rate of the water sprays that cool the hot steel must be recomputed for each change in the thickness or composition of the steel.

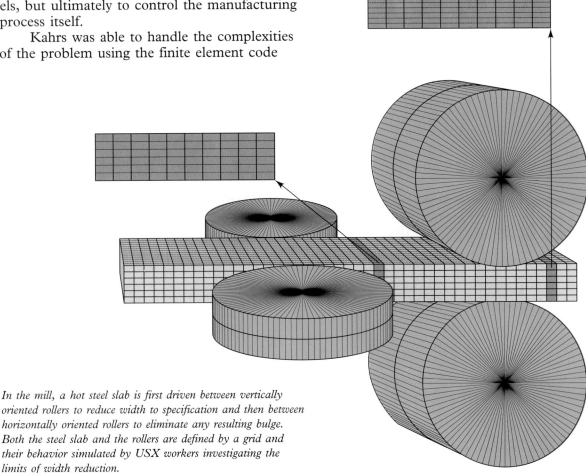

In the mill, a hot steel slab is first driven between vertically oriented rollers to reduce width to specification and then between horizontally oriented rollers to eliminate any resulting bulge. Both the steel slab and the rollers are defined by a grid and their behavior simulated by USX workers investigating the limits of width reduction.

The simulations have allowed USX to identify the steel dimensions that can be achieved without damaging the very expensive capital equipment in the mill. As a result, the company can manufacture a steel product that is much closer to the customer's specification and yet less expensive, because of reduced waste. As computers become faster and cheaper, the manufacturer will ultimately be able to embed this level of computational power directly in the milling equipment, so that a programmable and therefore more flexible milling factory will be possible.

Process Engineering: Food Canning and Sterilization

The U.S. food industry applies heat to over 100 billion cans of food each year in order to destroy microorganisms that cause spoilage and disease. A central issue for the canning industry is the time needed to process the food. One surviving spore of the bacterium *Clostridium botulinum* can replenish the population quickly enough to produce sufficient toxin to kill a person within hours of ingestion. Oversterilization, however, degrades both the taste and nutrient content of canned foods, and costs the food industry time and energy.

To demonstrate the utility of supercomputer simulations in food processing, Ashwini Kumar and Kenneth Swartzel of the Food Science Department at North Carolina State University modeled the heating of a cylindrical can containing a liquid food and a single meatball. The meatball is fixed on the axis of the can one-third of the distance from the can's bottom. The positioning of the meatball on the axis gives the contents of the can an axisymmetrical geometry, and the assumption of axisymmetry simplifies the problem from three spatial dimensions to two. The can has a temperature of 40°C at the start of the simulation, when the can's walls are exposed to the commonly used sterilization

temperature of 121°C. The liquid food then heats up as the liquid transports the heat from the walls to the can's interior by flowing and mixing in the natural process of convection. At the same time, the inside of the meatball heats up by conduction, the process through which energy is transferred through solid materials.

Kumar and Swartzel solved the equations governing the food's behavior using the commercial fluid dynamics application code FIDAP (the same program used by Dantzig to cast steel) on a Cray-2 at the Minnesota Supercom-

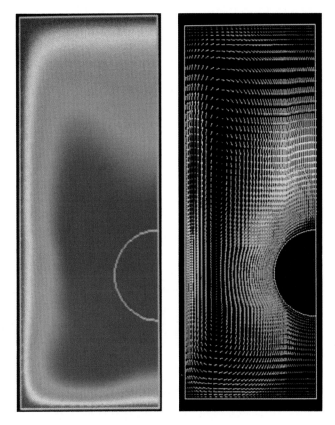

Left: *Three minutes after exposure to heat, the temperature in a can of liquid containing a meatball ranges from 40°C (blue) in the interior to 121°C (red) near the edge.* Right: *A field of arrows depicts the motions of food inside the can. The length and orientation of the arrows show the direction of motion, and the color indicates speed, red for high velocity and blue for low.*

puter Center and a Cray Y-MP at the North Carolina Supercomputing Center. They found that heated liquid near the wall of the can rose while cold liquid in the core moved downward, around the meatball, and then outward toward the wall near the bottom of the can. The velocity of the liquid peaked after about three minutes of heating. The zone to heat most slowly determines the duration of heating needed to ensure the inactivation of microbial organisms; this zone was found to be just below the center of the meatball. Computing the time required to heat the slowest-heating component allows canners to tailor the sterilization process to the specific densities of the food product in question.

These methods are appropriate for modeling a can that sits passively as it soaks up heat. However, nearly half of all cans processed worldwide execute complex trajectories while absorbing heat inside rotary pressure sterilizers of the kind manufactured by the FMC Corporation. These devices continuously move cans through a spiral track at rates as high as 800 cans per minute. The movement of the cans stirs their contents in order to improve heat transfer and lessen the resident time in the sterilizer. Unfortunately, it is virtually impossible to measure the velocity or temperature of the contents of the can once it is inside the sterilizer.

Turning to supercomputers, researchers of the FMC Corporation developed an interactive graphics interface to a custom-designed program that solves the equations of motion of the can and its contents. For instance, when the can is fed into the machine it "leaps" across a small space and lands on its seat on the sterilizer reel. Ultrathin metal cans can easily be dented if they hit with too much force. Using the simulation package, the engineer can adjust all of the design parameters such as the geometry and the rate of feed to achieve the fastest processing rate that avoids unwanted damage. FMC engineers were able to run their program on the Cray Research supercomputer at the National Center for Supercomputing Applications in Urbana, Illinois, while "turning the knobs" on their networked workstation at the Corporate Technology Center in San Jose, California.

Cans feed into an FMC rotary sterilizer at the right and emerge on the left.

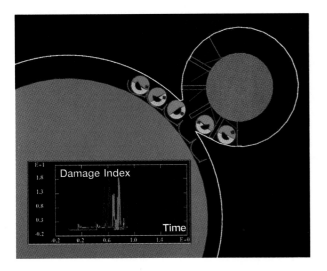

A can jumps into its seat in the sterilizer reel in this visualization of an operating rotary sterilizer. The graph plots the normalized contact stress as an index of damage to the cans potentially sustained at the current settings on the processing equipment.

Modeling Turbulence

Slow-moving fluids such as those encountered in the processing industry have flows that are fairly simple. Flows become much more complex for air or other fluids moving at higher speeds, particularly when the physical viscosity of the medium is important. To account for the effects of viscosity, one needs to solve the Navier-Stokes equations, which can describe turbulent flows. The ratio of the fluid's inertial resistance to acceleration to the fluid's viscosity is a constant termed the Reynold's number. A flow with a low value of the Reynold's number will be smooth and laminar, while a flow with a high value will be turbulent. In a turbulent flow, the bulk kinetic energy of the flow is degraded into smaller and smaller vortices. Experiments and observations have shown that large-scale vortices can also form and be sustained in such flows: the Great Red Spot in Jupiter's turbulent atmosphere is an example.

For turbulent flows whose Reynold's number is not too high (a few thousand), today's supercomputers can solve for fully three-dimensional turbulent flows using a modification of finite differencing. In a computational problem solved over a uniformly spaced grid, the size of the grid zones limits the resolution of small eddies in the flow. Thus, high resolution is needed to model the turbulence that always develops when a fluid like air flows over a surface, perhaps an airplane wing or the inside of a heating duct. The velocity of the gas must vanish at the surface, so there is a very thin "boundary layer" within which the velocity climbs from zero to the mean value in the flow. In air flow through a duct, for instance, this transition region is about 1 millimeter in width. The rapid change in velocity across this region of space can seed the vortices seen in turbulent flows.

To capture the details of the flow in a boundary layer, a simulation must have a finer grid near the surface than in the middle of the flow. One method for refining the grid near a surface is a modification of finite differencing termed a spectral method. It is a generalization of the time-honored method called Fourier analysis, which expresses physical variables as the sums of sine and cosine functions multiplied by unknown coefficients. In spectral methods one constructs sums from more complicated exact mathematical functions, then uses them to refine the grid near the surface defining the boundary layer.

Simulations of three-dimensional turbulence using spectral techniques are among some of the most time-consuming engineering problems run on supercomputers. The image on the next page of the fluctuations in the velocity and temperature fields in a gas flow between two horizontal walls took 400 hours to compute on a Cray-2 supercomputer. The image is from a study of turbulence production performed by S. L. Lyons and T. J. Hanratty of the University of Illinois at Urbana-Champaign and J. B. McLaughlin of Clarkson University. Periodic boundary conditions, like those Woodward used

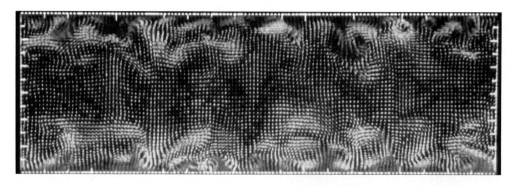

In this simulation of gas flow between two horizontal walls, eddies generate large coherent structures that are stretched downstream by the bulk flow. The direction of flow is toward the reader. The length and orientation of the lines show the local velocity of the fluid flow, and the lines are colored on a rainbow scale from blue to red according to the local temperature.

in Chapter 1, are imposed on the right and left so that no boundary effects come from those regions.

The channel was heated at the bottom and cooled at the top; the turbulent mixing of gas layers of differing warmth creates fluctuations in temperature that serve to trace the structure of the vortices. The exact functions used to define the grid in the vertical direction are called Chebyshev polynomials, while sines and cosines are used to define the grid horizontally and in the direction of the flow, outward from the page. The vertical resolution improves by a factor of 40 as one moves from the center of the channel to the top or bottom wall, thus permitting the accurate resolution of eddy generation in the boundary layer.

Such simulations have revealed that large-scale coherent structures form in turbulent flows near walls. In another supercomputer simulation carried out by P. R. Spalart of NASA Ames Research Center on Cray Research supercomputers and analyzed by his colleagues S. J. Kline of Stanford University and S. K. Robinson of the NASA Langley Research Center, low-pressure "vortex tornados" have been discovered near the wall. This high-resolution simulation performed over 9 million grid zones computed a flow of Reynold's number 670 in the turbulent

boundary layer near a flat plate. The characteristic "arch" shape is also seen in experiments and is intimately associated with turbulence generation. Insights from such fundamental engineering simulations can lead to better practical designs to improve fuel efficiency, lower noise generation in engines, and reduce frictional drag in air duct systems in homes and office buildings.

Air Flow in and around Automobiles

The automobile industry was the first group of manufacturers to exploit supercomputer modeling extensively. During the 1980s just about every automobile manufacturer in the world bought its own supercomputer. The availability of these supercomputers in turn stimulated the development of a great deal of software. As a result, by today, virtually every aspect of a car can be simulated: air flow over and inside the car, its vibration and noise, the structure and operation of all its subcomponents, and its handling of a wide range of safety tests. Thus designers and testers can work directly on a digital representation of the car. Such models increasingly guide manufacturing as well: for example, they can show how dies interact with metal sheet deformations or how to minimize the steel

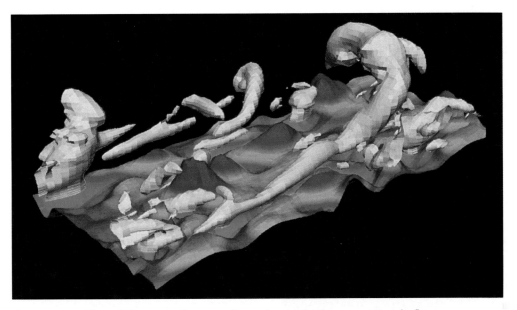

Computer graphics techniques reveal pressure fluctuations and coherent structures in flow through a turbulent boundary layer near a flat plate. The height and color of the floor give the pressure on the wall. Peaks are magenta and valleys blue, while green represents the mean pressure. The white contour surfaces represent constant values of low pressure. Visualizations of the velocity seen in cross sections through the surfaces reveal strong vorticity within flow. Flow is from left to right in this NASA simulation.

lost in casting. By the end of this decade we can expect computers to coordinate the entire process of producing an automobile.

Consumers and government regulators are pushing automobile manufacturers to build cars that are more fuel efficient. Of the many factors that influence efficiency, some of the most important are the flow and combustion of the air-gasoline mixture in the engine, and the aerodynamic drag exerted on the car's body. Thus detailed, realistic simulations using computational fluid dynamics can be of great value to engineers designing a new model.

In order to model the flow of air and fuel through the intake system of an automobile engine, which consists of an inlet port, intake valve, and combustion chamber, one needs a very flexible code, one that is able to handle flows in a variety of physical states of temperature and compression and that can also achieve a high degree of geometrical flexibility through the use of unstructured finite element meshes. A good example of such a code is STAR-CD, developed by the London-based firm Computational Dynamics.

One simulation (next page) ran on this code, using over 200,000 finite element nodes, revealed the complex behavior of the air-fuel mixture moving through the system, including the characteristics of the flow at the valve orifice, which can dramatically affect the combustion performance. For example, the opening of the inlet valve causes a pressure drop that significantly affects the engine's ability to "breathe." When compared with measured values of the pressure drops, the code was found to agree within a few percent. Using such verified codes, the engineers designing the next generation of automobiles are better able to increase an engine's power while lowering its fuel requirements.

The velocity field of the flow adjacent to surfaces in the inlet port and cylinder of an automobile engine. Red is high velocity, and blue is low.

Modeling ignition and combustion in the cylinders is a challenge that takes supercomputing application programs to the edge of their capabilities. Not only does the program have to resolve the air flow and locate the zone where burning occurs, but it must correctly resolve the fuel breaking up into a spray of droplets. Then it must compute the chemical reactions undergone by the burning fuel under these conditions of flow and spray, and the influence of that chemistry in turn on the flow development. The goal of these simulations, which are being tailored to run on either vector multiprocessors or massively parallel supercomputers, is to design engines that burn fuel more efficiently, both in order to improve gas mileage and in order to reduce unwanted pollution.

A good example is work taking place at Los Alamos National Laboratory, where researchers in the Fluid Dynamics Group T-3 are simulating a Direct-Injection Stratified Charge engine.

The air in this engine rotates so that as the fuel is injected it swirls around as the droplets break up. The spray evaporates to a gaseous fuel-air mixture that ignites when a spark is applied. The simulation computes the fuel burning, then visualizes the combustion by showing the increase in temperature as the engine runs through its cycle.

Modeling air flows can also help fuel economy by suggesting how the shape of a car may best be altered to reduce air drag. A unique approach to computing automobile aerodynamics is taken by Kunio Kuwahara, who directs the Institute of Computational Fluid Dynamics in Tokyo. This organization is located in Kuwahara's house and garage, where he keeps one each of the major Japanese supercomputers from NEC, Fujitsu, and Hitachi. Kuwahara has developed methods for solving the Navier-Stokes equations, which he uses to compute the flow around a complete automobile. For instance, in the late 1980s, Kuwahara and Susumu Shirayama tackled the particularly difficult problem of computing the flow around a car with tires.

Aerodynamic simulations of automobiles often omit the tires, because the contact of the tires with the road makes it difficult to generate a grid that covers the tires accurately and also covers the air around the car. Yet tires are a major source of the turbulent drag on the automobile. Shirayama and Kuwahara invented a sophisticated grid transformation technique to create a mesh system of $150 \times 70 \times 75$ grid points that covered the tires and the surrounding air. Using finite differencing methods, Shirayama and Kuwahara solved the Navier-Stokes equations for cars both with and without tires.

Their simulations clearly elucidated the effects of tires. The pressure on the front fender is greater when tires are present because a downwash precedes the front tires, followed by an upwash immediately behind the tires. The addition of tires also gives rise to a broad deflection of air flow around the lower part of the car, creating numerous longitudinal vortices that per-

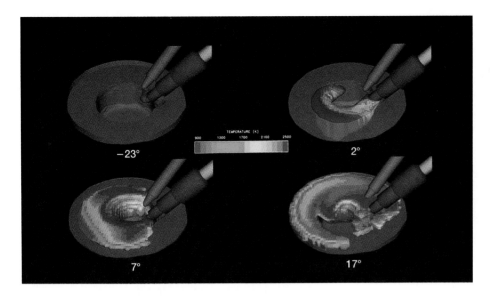

Rotating air swirls fuel around until the spark ignites the mixture. The color indicates the temperature in the fuel-air mixture; blue is coldest and red hottest. The number below each image indicates the increase in crank angle in degrees.

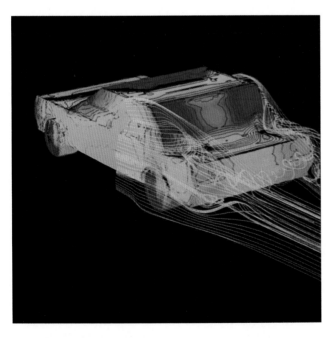

In this simulation of air flow around a car with tires, white lines trace the paths followed by particles released from various locations around the car. The swirls near the back of the car reveal turbulence. The color of the car represents surface pressure.

vade this separated region. However, air flow over the upper region of a car is relatively unaffected by the presence of tires, as one might expect.

Increasing Automobile Safety

After studying the gross features of the car shape, engineers make more detailed computations of individual subcomponents. For instance, the tires that so complicate the air flow computations are themselves extremely complex engineering marvels. The average load on a tire is 50 times its own weight; at high speed any given piece of the tire exterior is in contact with the road for only a few thousands of a second; and hard braking drives temperatures to over 200°C. Most of us never think about the rigors expected of a tire but assume the tire will perform its task for tens of thousands of miles.

Tire designers, such as A. Guillet at the French company Michelin, have developed elaborate engineering codes that can take into ac-

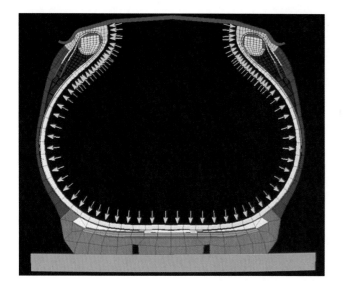

In this simulation of a cross section through a Michelin tire, yellow arrows indicate the force of the internal air pressure on the tire. Colors represent the different materials used in the tire construction. The computer program must take into account the material characteristics of each component.

count the properties of the materials in tires, the changes in these properties with temperature and speed, the interaction of the tire with the car and the road, and the deformation of the tire under different conditions. By using 60,000 finite elements on a Cray X-MP, Guillet's group was able to simulate the stresses on a tire and its components within a few hours. This computational tool enables Michelin to test proposed changes immediately, thus accelerating the design process.

Perhaps the single most important class of automotive simulations are those that increase the safety of passengers during an accident. Until the mid-1980s, an automobile company could ensure that its new models met legal safety requirements only by performing laboratory crashes on instrumented production cars, at a cost of hundreds of thousands of dollars per test. Since the advent of finite element codes for crash analysis, most of the world's automotive companies have supplemented their laboratory crash tests with supercomputer simulations.

The advantages of simulated crashes are many. One or more runs can be made overnight, a much faster turnaround than can be achieved with physical tests. Consequently a car model can be tested repeatedly during its design. Unlike physical tests, the simulation provides data on the stresses and accelerations at all points in the vehicle. And finally, the simulated crash is much cheaper. As a result, simulated car crashes are the only practical means of estimating whether a new car will meet the increasingly stringent government safety standards as its design progresses from the drawing board to the assembly line.

Two-thirds of all automobile crashes worldwide are head-on collisions. In three-quarters of these, the fronts of the cars overlap only partially. Mercedes-Benz of Sindelfingen, Germany, has performed extensive tests of this most frequent of accidents. K. Gruber, T. Frank, and their colleagues at Mercedes-Benz can compute the results of a crash in fifteen hours on a Cray Y-MP, using the DYNA3D finite element code, which can handle large displacements and rotations. Their model used 20,000 finite elements to simulate the 55 kilometer per hour impact shown in the image on the facing page. As can be seen, the damage from the simulated crash agrees very well with the damage inflicted on a test car by an actual crash. The results of the tests have led to a number of structural modifications that have improved the passive safety of the passengers.

Although frontal impacts are the most common, side impacts still account for about 30 percent of the deaths in automobile accidents. To lower fatalities, the National Highway Traffic Safety Administration in the United States and its European counterparts are formulating legislation that will specify the maximum acceleration that physical crash-test dummies should experience in the rib cage, lower spine, and pelvis during side impacts.

Aruna Tilakasiri at Ford Motor Company in England and his German consultant Paul Du Bois tackled the problem by incorporating finite element dummies, developed at the Ford Research and Engineering Center in Detroit, in their simulation of a side impact. Two finite element dummies were placed on the impact side of the car, one in the front seat and one in the rear seat. The dummies' heads, rib cages, lower spines, pelvises, and legs were modeled as independent rigid bodies defined by finite elements. These body parts were joined by simulated springs and dampers to form the complete dummies. Using 9000 elements in a crash code called RADIOSS, Tilakasiri and Du Bois were able to model a Ford Sierra, a medium-sized family hatchback. Their simulation modeled the effects of a light truck traveling at a speed of about 30 mph impacting the side of the Ford Sierra moving at about 15 mph. The simulation was carried out remotely on a three-processor Cray Y-MP at Ford in Detroit.

As seen on the next page, the dummies are hit by the inside of the door panel between 10 and 20 milliseconds after the truck strikes the outside of the car. Roughly 20 milliseconds after that, the rib cage, lower spine, and pelvis experience peak sidewise accelerations of 50, 60, and 100 times the normal vertical acceleration due to gravity (g). The entire collision is over in less than one-tenth of a second! These detailed simulations not only will be used to re-engineer the car frame, but will also provide critical data for the development of the safety legislation itself.

This and similar research has demonstrated that engineers can perform detailed studies of a car's design years before a prototype is available for testing. As the car moves toward production, changes are constantly being made by groups trying to meet the conflicting requirements of efficient gas mileage, rigorous safety measures, reasonable manufacturing costs, and attractive style. Working from models based on the evolving design database that defines the car, these groups will be able to solve problems quickly to meet their objectives within the same time

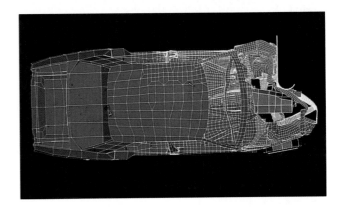

Top: *A Mercedes-Benz automobile was rammed into a test wall at 55 kilometers per hour, creating a partial overlap impact.* Bottom: *A fifteen-hour simulation of the same event produces a very similar image, evidence of the power of supercomputer simulation in improving automobile safety.*

frame. The ultimate fruit of this supercomputer-aided process is a safer, more fuel efficient, less polluting means of transportation for the consumer and a more cost-effective, competitive product for the manufacturer.

Engineering Safer Nuclear Reactors

The economic health of many nations as well as the environmental health of our planet will clearly benefit from forms of energy production

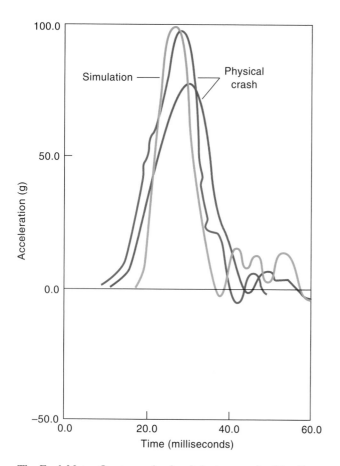

The Ford Motor Company simulated the impact of a 30-mile-per-hour collision on the side of an automobile moving forward at 15 miles per hour. The accelerations experienced by the front pelvises of instrumented dummies in two physical car crashes (blue and red lines) agree closely with the results of a supercomputer simulation (green line).

that do not consume fossil fuels. Two technologically complex approaches to creating alternative fuels involve tapping the energy source in the nucleus of atoms. Nuclear theory tells us that energy can be liberated if very heavy nuclei are fissioned into smaller nuclei or conversely if light nuclei are fused together to form heavier nuclei. Fission forms the basis for today's nuclear reactors, and fusion may become an im-

portant energy source by the middle of the next century.

The success of both energy sources requires the resolution of long-outstanding problems concerning the safety of the designs, the long-term disposal of the radioactive wastes, and the affordability of the installations. All three of these legitimate concerns are being studied computationally. Here we shall briefly touch only on the use of supercomputers to help better understand the basic engineering of such reactors.

A self-sustaining fission reaction continuously generates heat, which is absorbed by liquid running through the core area. The heated liquid then is used to generate electricity. If, for some reason, the coolant is prevented from reaching the core of the reactor, the heat accumulates until it could conceivably set off a core meltdown. Just such a malfunction of the fission nuclear reactor at Three Mile Island in 1979 focused public attention on the safety of reactors. Because of the design of most operating reactors in the United States, "loss of coolant" is the most feared accident.

Nuclear engineers have been exploring a number of alternative designs that will not allow the core to melt down if the coolant is lost. These designs incorporate passive safety features that will work reliably independently of operator actions or of complicated active systems. One such design is called the Modular High Temperature Gas-Cooled Reactor (MHTGR), being developed with Department of Energy funding by the General Atomics Corporation, the administrators of the San Diego Supercomputer Center.

The MHTGR reactor will be placed underground so that if the core starts to overheat, passive heat convection will transmit the heat to the external earth, which acts as a massive heat sink. The coolant will be helium, an inert gas that cannot interact with core materials to create an explosion as water can. In addition, the reactor has an annular core geometry surrounded by graphite inner and outer neutron reflectors to concentrate the neutron flux in the active core

Temperatures inside the Modular High Temperature Gas-Cooled Reactor are visualized as colors, from deep red for high temperatures to dark blue for low.

region. Simulations on a Cray Y-MP reproduce the temperatures found throughout the core in response to a variety of possible accidents. Not only is this method critical to the design process, but the resulting animations may assist the slow regaining of the public's trust in the nuclear industry.

Taming the Hydrogen Bomb

In the fifteen million-degree inferno at the Sun's core, hydrogen nuclei are fused together to produce helium nuclei. During this union, mass is converted into energy that eventually makes its way to the Sun's surface and escapes into space as sunlight. Laboratory temperatures on the order of 200 million degrees centigrade are required to increase the speed of hydrogen nuclei enough to overcome their mutually repulsive electric fields, so that the nuclei may approach close enough to fuse. Scientists and engineers learned how to bring a bit of the Sun's core down to Earth in the 1950s when they invented the hydrogen bomb. At the same time an international search began for a way to create peaceful energy from hydrogen fusion.

The ultrahigh temperatures pose special problems for engineers attempting to produce controlled thermonuclear fusion in the laboratory. No physical container can endure such temperatures. Instead researchers use powerful magnetic fields in an attempt to confine the hot gas, which is actually a mixture of free nuclei and electrons called a plasma. The motions of charged particles are strongly restricted by the geometry of a magnetic field, and so "magnetic bottles" are designed to hold the plasma as temperatures rise toward the ignition point for fusion.

A major stumbling block on the road to a thermonuclear fusion reactor is the fact that many sorts of instabilities can develop in the plasma that destroy the magnetic confinement. The most violent instabilities arise in a matter of microseconds. An efficient and cost-effective method of studying such instabilities is to use a supercomputer to solve the three-dimensional magnetohydrodynamic equations that describe the motions of a magnetically confined plasma. In this model, the charged particles are required to move only along the magnetic field lines. The goal of these computations is to find a reactor design that allows a high-density plasma to maintain a hundred-million-degree temperature for at least one second.

One of the most successful designs for a fusion reactor, called a tokamak from the Russian term meaning "strong current," uses magnetic fields to push the plasma away from the walls of a toroidal or doughnut-shaped vacuum vessel. Electromagnets wrapped around the vessel produce a toroidal magnetic field that follows the geometry of the vessel. Transformers around the vessel induce a strong electric current that flows through the plasma, thereby producing a second magnetic field that acts together with the externally applied magnetic field to confine the plasma. During the 1980s, the stabilities of tokamak configurations were analyzed by extensive computations that could be carried out in two spatial dimensions because a torus is axially symmetric. The results were employed in the design of the Tokamak Fusion Test Reactor at Princeton and the Joint European Torus in England.

An alternative successful design, called a stellarator, does not include an induced current, which can give rise to instabilities. Rather, it relies entirely on electromagnets to confine the plasma. The stellarator is inherently three dimensional, requiring a more complex simulation.

In 1988, an international team of researchers from the Centre de Recherches en Physique des Plasmas in Lausanne, Switzerland; the Max Planck Institute for Plasma Physics in Munich, Germany; the National Magnetic Fusion Energy Computer Center at Livermore, California; and the Cray Research Corporation tackled the formidable chore of developing a three-dimensional program, called Terpsichore, for evaluating the stability of several fusion designs. To ensure the efficiency of Terpsichore, the developers carefully chose numerical methods well suited for implementation on parallel vector computers. Indeed, the advanced numerical methods employed in Terpsichore yield a sustained computing speed of 1.7 gigaflops on an eight-processor Cray Y-MP, 65 percent of the maximum possible speed!

The first step in a plasma stability analysis is to compute a solution to the equations of magnetohydrodynamics for the case when the plasma is in a state of static equilibrium. The

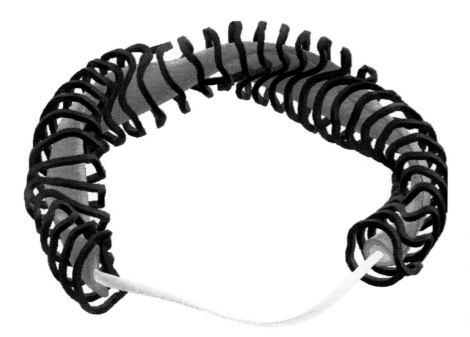

A simulation of the proposed Wendelstein VII-X stellarator shows the complex shape of the plasma as it twists within the coils. The coils that generate the confining magnetic field are shown in grey, the plasma surface in red, and a magnetic field line in green.

results are fed into Terpsichore, which then perturbs the equilibrium solution and searches for unstable modes, in which the plasma literally explodes in only a few microseconds. The ultimate goal is to find configurations of magnetic fields and plasmas that do not have unstable modes. The program has also been used to simulate a proposed design called the Helias configuration, which is being considered for the Wendelstein VII-X stellarator device, the next stellarator experiment at the Max Planck Institute for Plasma Physics.

The configuration can now be modified on the supercomputer and recomputed until the best configuration is found. Compare this approach with having to build each configuration in the laboratory and then finding out it would not work as hoped. The careful interplay between simulation and experiment will continue for decades to come, as the world continues its quest for safe, economic, and nonpolluting energy.

Structural Engineering for Safety

Two solid-rocket boosters provide most of the thrust that lifts the Space Shuttle off the launch pad. The boosters are made in four sections at the Utah plant of Morton Thiokol and then joined together with "field joints" at the Kennedy Space Center in Florida. After propelling the shuttle to an altitude of about 30 miles and a speed of Mach 4, the boosters separate from the external tank and parachute back to Earth for refurbishing and reuse. In a historic disaster seventy-three seconds after liftoff on January 28, 1986, the Space Shuttle *Challenger* exploded, killing the crew and bringing the U.S. space program to a sudden halt. The culprit was a failed O-ring seal on one of the field joints at the rear of a solid rocket booster. Flames leaking through the failed seal ignited the external fuel tank, causing it to explode.

Much of the Space Shuttle had been designed in the late 1960s and early 1970s using computers whose memory and speed were roughly comparable to those of ordinary personal computers today. Because of the low computing power, the original analyses of the booster rocket design were restricted to simplified two-dimensional axisymmetric models. Shortly after the *Challenger* disaster, the task of using supercomputers to redesign the solid rocket boosters was taken up by Morton Thiokol and NASA staff at the George C. Marshall Space Flight Center in Huntsville, Alabama. The large increase in computing power provided by acquistion of a Cray X-MP permitted a fully three-dimensional structural analysis. Every facet of the booster design was scrutinized, and special attention was given to the O-ring seals. Three commercial finite element programs, MSC/NASTRAN, ABAQUS, and ANSYS, were pushed to their limits in the redesign process.

In the original booster design, both the field joints and the joints already mated in the factory had the same basic structure consisting of a tang and clevis bolted together by 180 one-inch-diameter pins. Two groves on the interior lip of the clevis each held an O-ring. Nathan Christensen and his colleagues at Morton Thiokol performed a complete finite element analysis of the response of the joint to extreme loading. The model ran for nearly 19 hours on the Cray X-MP, producing color-coded contour plots of the plastic stress on the joint shown on the next page. The analysis indicated that the factory joints performed well, because they have continuous layers of liner and propellant cast over them. Although the factory joints are therefore not subjected to direct pressure by the hot internal gases, the field joints are. The analysis showed that the mating surfaces had too much freedom of movement to maintain continuous sealing contact during extremes of pressure and temperature.

While only minor changes were made to the factory joints, substantial modifications were

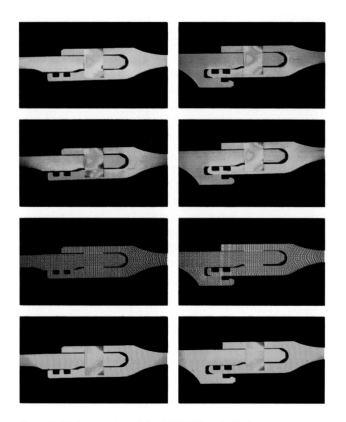

Stress is higher on the original field joint (left) than on its redesigned counterpart (right) in this sequence of images showing, from top to bottom, plastic stress at 0.2-second intervals during ignition of the solid rocket booster. The magnitude of stress is indicated by a rainbow scale of colors ranging from blue for low stress to red for high stress. The two images at the bottom show the finite element mesh.

necessary in the field joints. After running finite element models of more than 25 different joint configurations, the Morton Thiokol engineers concluded that best performance was achieved by adding a retainer lip to the tang along with a third O-ring seal, as shown in the images on the left. The engineers also increased the size of the two grooves in the clevis and their O-rings. The pins that secure the joints were lengthened, and retention bands were placed around the booster

to hold the pins in place. Other additions include a joint heater, insulation, shims, and weather seals.

The desire to resume shuttle flights as quickly as possible was a primary motivation in using supercomputers to redesign the shuttle booster. Only by using thousands of CPU hours on a machine like the Cray X-MP could engineers hope to achieve a timely resumption of launches. As a result, the redesigned boosters are one of the most analyzed pieces of hardware in history.

The Aerodynamics of Space Vehicles

The aerospace industry is one of the most sophisticated in its use of supercomputing to support engineering design and analysis. Virtually every type of commercial, military, and space vehicle is extensively investigated using supercomputer simulation. We can hope only to touch the surface of this field here. Indeed, much of the work is classified, as the sudden success of stealth fighters in Desert Storm testified to. Nonetheless, by studying a few of the public examples, we can gain an idea of what supercomputing will bring to other industries that have only started to explore the use of high-performance computing.

Engineers designing airplanes or space vehicles use expensive wind tunnels in order to study the details of air flow around an airfoil or fuselage, even though the equations describing such flow have been known since the nineteenth century. John von Neumann realized in the 1940s that supercomputers solving these air flow equations should be able to give the same answers that wind tunnels do. "The purpose of the experiment is not to verify a proposed theory but to replace computation from an unquestioned theory by direct measurement," von Neumann wrote. "Thus wind tunnels are used at present, at least in large part, as computing devices of the so-called analog type to integrate the nonlinear partial differential equations of fluid dynamics.''

In the fifty years since von Neumann's visionary statement, supercomputers have finally gained the speed and memory needed to begin to fulfill his vision. Wind tunnels have not been replaced; rather, they are used alongside supercomputer simulations to increase our understanding of the complexity of air flows. For instance, consider a simulation performed by Peter Buning, I. T. Chiu, S. Obayashi, Yehia Rizk, and Joseph Steger at NASA Ames Research Center in California. It solves the Navier-Stokes equations to model the behavior of air flowing past the Space Shuttle as it rockets into orbit. The simulation required about five hours on a Cray-2, and as the image on this page shows, the closeness of the simulated values to measured values vindicates von Neumann's vision.

The future of human spaceflight will include an Earth-orbiting space station, permanent bases on the Moon, and missions to Mars. To achieve these goals, we must be able to transport people and supplies to and from Earth-orbit more efficiently and economically than is now possible with the Space Shuttle. Scientists at NASA as well as in private industry are therefore designing a new vehicle, called the National Aerospace Plane (NASP), that will be the workhorse of the space program early in the twenty-first century.

The National Aerospace Plane is an air-breathing, hypersonic jet airplane that will have sufficient thrust to reach orbit from a horizontal takeoff. The primary advantage of an air-breathing system is that the vehicle will not need to carry an oxidizer. Traditional rocket propulsion systems, like those of the Space Shuttle, must haul oxygen-rich chemicals with which to burn their fuel. By utilizing oxygen in the upper atmosphere instead of carrying an oxidizer, the NASP will be able to transport greater payloads to and from orbit.

A supersonic combustion RAM jet, often termed a scramjet, is the only type of engine capable of extracting an amount of oxygen from the sparse upper atmosphere sufficient to keep the plane flying. In a scramjet, the air flow remains supersonic all the way through the engine. Hydrogen is the only fuel that burns quickly enough to produce adequate thrust during the brief moments that the air-fuel mixture is in the engine. Using a scramjet and hydrogen fuel, the NASP will be able to fly at altitudes up to 200,000 feet and achieve a peak speed of Mach 25. In the mid-1980s, university, government, and industry researchers began developing Navier-Stokes codes that would solve the com-

White represents the highest pressure and blue the lowest in this map of the surface pressure on the Space Shuttle. The right half of the image shows the actual values measured by sensors, while the left half shows the predictions of a supercomputer simulation.

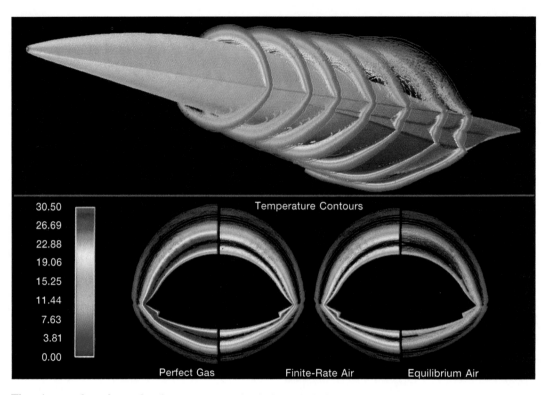

30.50
26.69
22.88
19.06
15.25
11.44
7.63
3.81
0.00

Temperature Contours

Perfect Gas   Finite-Rate Air   Equilibrium Air

These images show the results of supercomputer simulations of air flowing over the NASP flying at an altitude of 61 kilometers with a speed of Mach 25. Top: *Computation of the atomic oxygen level shows that more oxygen has been dissociated into atoms (red) close to the wings, where the temperature is higher.* Bottom: *The temperature predicted near the body of the plane varies a great deal depending on whether the air is modeled as a perfect gas with no chemistry, a gas with chemistry in equilibrium, or a gas with chemical reactions proceeding as the air flows over the body of the plane.*

plete flow field, both over the body and through the engine, for a NASP-like vehicle. Ultimately, the designers of the NASP will assign parameters to various features of the vehicle, and successive simulations on supercomputers will reveal the best choice of the parameters.

Molecules of atmospheric nitrogen and oxygen become excited and dissociated as they pass through the shock that envelops the NASP's forebody, absorbing energy that would otherwise heat the surface of the plane. During the late 1980s, scientists began modeling these chemical processes to determine their detailed

effects on air flow. For low-altitude flight in moderately dense air, it is generally assumed that the gas is in chemical equilibrium everywhere in the flow field. This assumption allows one to simply take conventional Navier-Stokes codes and add a model of the gas chemistry at equilibrium. At high altitudes, however, the gas leaves equilibrium and chemical reactions proceed at the same rate as the air flows over the vehicle.

Thomas Edwards and Jolen Flores at NASA Ames Research Center in California incorporated chemical effects into NASA's

Navier-Stokes code by including seven reactions between six species of atoms and molecules (N_2, O_2, NO, N, O, and NO^+) and free electrons. The chemical reactions reduce the temperature of the gas streaming past the vehicle because the excitation, dissociation, and ionization of the molecules and atoms absorb energy. As air tra- verses the shock, oxygen molecules begin to dis- sociate into oxygen atoms. However, the propor- tion of oxygen molecules dissociated is small, because the temperature behind the shock is not extremely high. The full complexities of these reactions and their effects on the NASP design will require teraflop supercomputers to unravel.

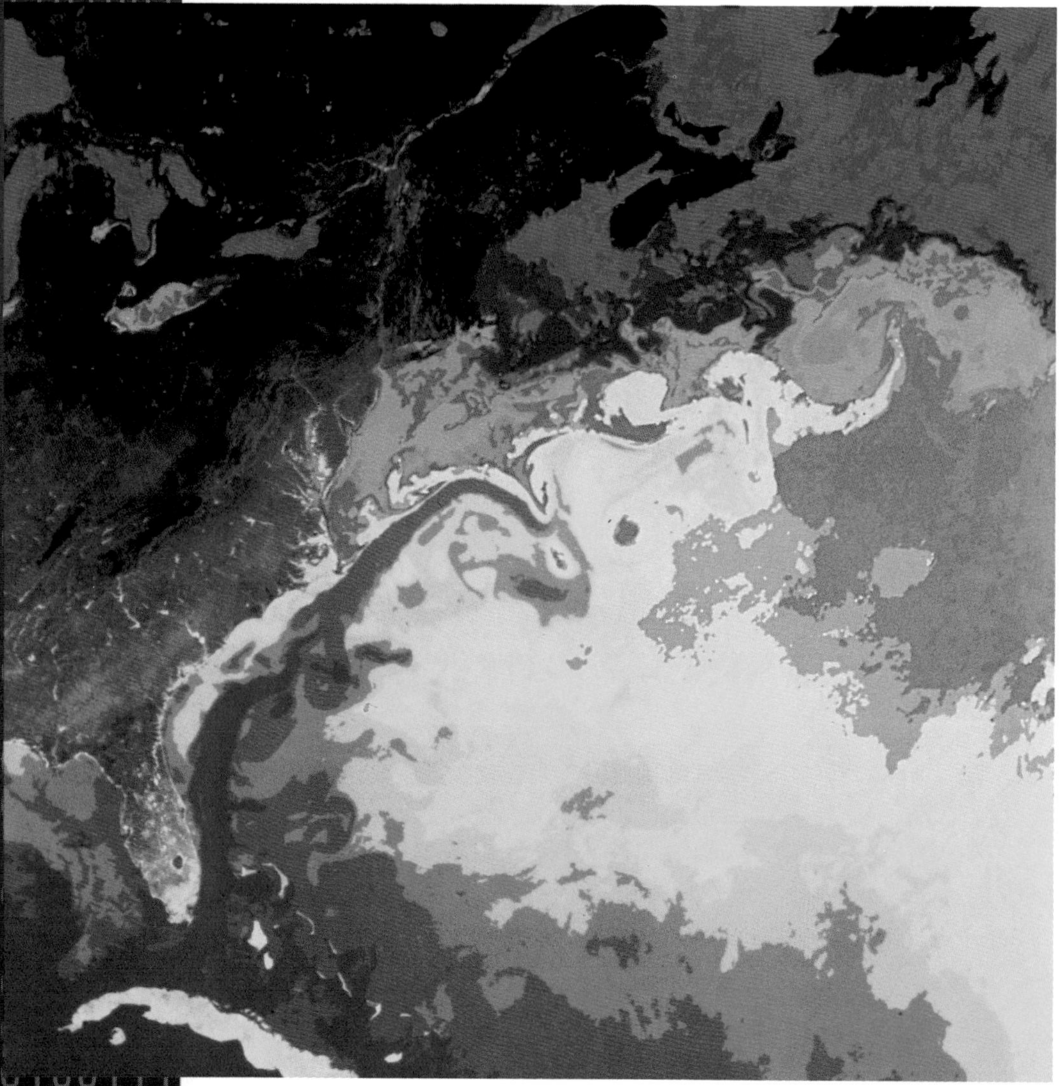

Our Dynamic Planet

7

The Earth is an unusual planet in that three-quarters of its surface is covered by liquid water. The continual evaporation of water vapor into the atmosphere creates a "greenhouse effect" that keeps the Earth's surface much warmer than it would otherwise be at its distance from the Sun. This combination of water and warmth, unique in the solar system, allowed life to evolve shortly after the formation of the Earth.

Two to three billion years ago, one-celled organisms developed photosynthesis and began to "poison" the atmosphere with free oxygen. As a result of this massive "pollution" episode, higher life forms evolved that used oxygen for metabolism. These organisms eventually became the plants and animals that moved from the sea onto the land and into the air. In the last few thousand years,

The warm waters of the Gulf Stream, in red, can be seen spinning off eddies as they flow northward through the northern Atlantic in this image compiled from data taken by the NOAA Advanced Very High Resolution Radiometer on the polar-orbiting NOAA weather satellite. These complex flows and their relationships to the climate and biosphere are now being simulated by supercomputing researchers.

one-millionth of the time that life has existed on Earth, human civilization has emerged. In the last hundred years, the waste products of these life forms have once again begun to alter the Earth system.

The realization that humanity must come to grips with these changes has proved a powerful motivator for scientific research. Many scientists are expending a great deal of effort building models of all aspects of the system formed by the Earth's oceans, atmosphere, land, and biosphere, including the energy, hydrologic, and chemical cycles that connect these components. Scientists have achieved considerable success building models of subsystems within the whole system; they have captured features such as the Earth's moving mantle, ocean currents, weather patterns, air pollution of cities, and wetland ecosystems. They have even integrated some of these subsystem models to create more comprehensive models. For instance, by combining models of the ocean, atmosphere, and land, scientists can model the physical climate system. Ultimately, these models will consider the detailed effects of the human activities that are now transforming our planet.

The Solid Earth

The Earth is still cooling off from its formation five billion years ago. After its condensation from the solar nebula, the heavier elements sank toward the core of the still molten planet. Since many of these elements, such as uranium or thorium, are radioactive and have half-lives measured in billions of years, they are still producing large quantities of heat. As a result, a steady flux of heat continues to travel from the core and surrounding mantle of the Earth to its surface even today. The thick, solid mantle is unable to carry this heat flux upward solely by conduction. Thus, the mantle material cycles slowly toward the surface and back, transporting the additional heat by convection.

The convective patterns within the mantle have caused the crust of the Earth to break up into plates that drift on top of the mantle. The drifting of the continents has been rearranging the Earth's surface for billions of years. When plates collide, mountain ranges are pushed up; when the edges of the plates are sucked down into the churning mantle, they generate earthquakes. This dramatic unifying picture of the solid Earth's upper layers, termed plate tectonics, has completely revolutionized the science of geology in the last three decades.

In the last five years, supercomputers have enabled researchers to simulate three-dimensional convection in the Earth's mantle. Using a Cray X-MP at the San Diego Supercomputer Center, Dave Bercovici of the University of Hawaii, Gerald Schubert of the University of California at Los Angeles, and Gary Glatzmaier of Los Alamos National Laboratory examined cases in which heat sources were only in the core, only in the mantle, or, in the most likely case, in both places. In all three cases they found the same pattern: heat is carried up from the core in plumes, and sheets of cold mantle material descend in regions surrounding the plumes. To discretize the equations, the scientists relied on spectral methods in space (as discussed in the turbulence simulation in Chapter 6) and finite differencing methods in time. Upward-welling plumes create "hot spots" such as the volcanic Hawaiian Islands or Iceland, while the descending sheets are associated with a worldwide system of trenches at the subduction zones of oceanic plates.

The petroleum and mineral resources accessible to human beings lie within a few kilometers of the Earth's surface. Many of these deposits are found in sedimentary basins that formed as layer after layer of debris silted in, containing particles that had settled out of rivers, accumulations of biological matter, and material transported by ocean and atmospheric currents. The evaporation of a body of water can later create thick salt beds, sections of which can rise through overlying rock to become salt

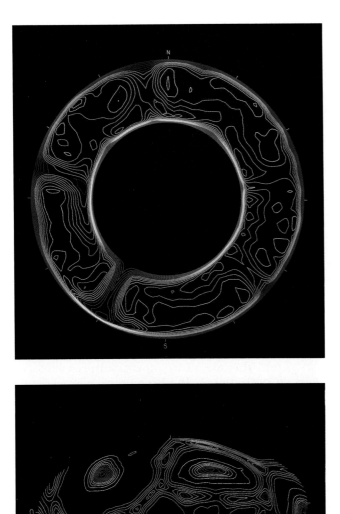

Hot mantle material (yellow/red) moves upward in a columnar plume, and cold mantle material (blue/purple) returns in sheets, in these simulations of mantle convection showing contour plots of the temperature content of the mantle (top) through a cross section of the Earth and its vertical velocity (bottom) over a spherical surface midway through the spherical shell above, shown in an equal area projection.

domes. Rocks deformed by the intruding domes create traps where petroleum and natural gas often accumulates.

Craig Bethke, a professor of geology at the University of Illinois at Urbana-Champaign, and Wendy Harrison of the Exxon Production Research Company in Houston, Texas, used an Alliant supercomputer in the mid-1980s to simulate the development of a sedimentary region in the Gulf of Mexico containing such petroleum traps. Their discipline, called hydrogeology, studies the very slow, but powerful flows of water and gases through porous rock formations. By diffusing from areas of high pressure to areas of low pressure, these subsurface fluid flows reduce the increasing pressures caused by compacting rocks. Areas of high geopressure form when new sediments are laid down faster than the subsurface can equilibrate or when impermeable layers prevent the normal diffusion of fluids from equalizing pressure. Drilling for oil or natural gas into a region of high pressure can result in an explosive oil well blowout.

Bethke and Harrison studied a cut 15 kilometers deep into the Earth's crust and 730 kilometers long, running north-south from east Texas into the Gulf of Mexico. The results from core samples and seismic profiles gave them the composition of the sediment and the rate of sediment deposition over the past 160 million years. These data were inserted into an equation governing flow through a deforming medium. The equation establishes the rate at which the pressure changes with time in response to the diffusion of the subsurface fluids, as modified by gravity, rock porosity, and heat flux. The solution to the equation gives the pressure at each depth along the cut as a function of time.

If equilibrium were maintained, the pressure should increase with depth at a rate of 100 atmospheres per kilometer. Instead, Bethke and Harrison found that in a large region the overpressure reaches 130 atmospheres per kilometer, and that in the last several million years a smaller region of even higher overpressure has formed, where the pressure is more than 160

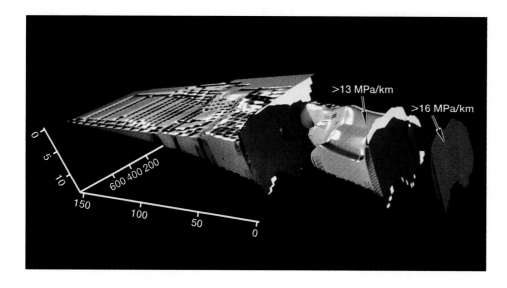

>13 MPa/km

>16 MPa/km

0
5
10
150
100
50
0
600 400 200

A three-dimensional rendering by Craig Upson of NCSA shows the development of geopressure in the Gulf basin over geological time scales. We are looking from the southern end of the cut inland toward east Texas. The blue region has a normal pressure gradient, the green surface encloses a region in which higher geopressures have developed, and the red surface encloses the region that has recently developed the highest overpressures. The enclosed surfaces have been exploded outward for easier viewing. The values of pressure gradient are given in megapascals per kilometer (1 MPa = 10 atmospheres).

atmospheres per kilometer. Apparently this most recent rise in pressure gradient occurred because the rapid uplift and erosion of the Rocky Mountains created a large amount of sediment, which was carried out to sea by the Mississippi River and dropped on the floor of the Gulf in a geologically short time. Simulations like Bethke and Harrison's can also show how water carries dissolved metals or petroleum to near the surface at the edges of a basin, thus explaining the otherwise seemingly random placement of mineral deposits and reservoirs.

In an attempt to locate more of these underground resources, all the major petroleum companies use highly computer intensive seismology techniques to image the rock layers near the surface. Mechanical "thumpers" or explosive charges send sound waves into the subsurface rock. The waves bounce off underground rock strata, and the returning echos are picked up by arrays of microphones. Each day, vast amounts of such unprocessed seismic data are collected all over the world. The petroleum industry has high-performance computers of all possible architectures turning tapes of this noise into images of the underground strata, which can reveal faults, salt domes, and other features known to

be associated with deposits of oil, natural gas, or minerals.

Until the advent of supercomputer-assisted seismic prospecting, the probability was high that prospectors would come up with a "dry" hole after spending millions of dollars on drilling. Today, the sophisticated computation and visualization of underground rock strata allow companies to avoid dry holes. Furthermore, because these analyses provide a detailed picture of an underground oil field, the oil company can make sure that all oil is extracted. The result is much more efficient management of the limited resources that support modern civilization.

The Ocean's Currents

The three great oceans of the world—the Pacific, the Atlantic, and the Indian—actually form one planetary-wide body of water, broken here and there with islands we call continents. Yet despite its surface extent of 20,000 kilometers from pole to pole, the world ocean is at its deepest only 8 kilometers. In contrast, the Earth's thick mantle extends 3000 kilometers in

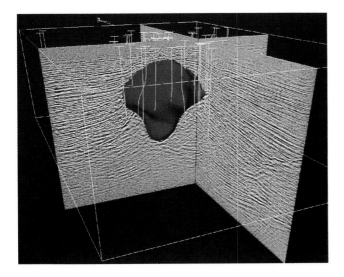

The three-dimensional striped volume shows rock strata surrounding a salt dome (red) in this image based on seismic recordings taken under an oil field by a petroleum company. The yellow lines show existing oil wells.

depth. All life on Earth depends on the thin film of water and the equally thin film of air covering it, because the interaction of the two layers controls the Earth's life-sustaining climate.

The large difference in temperature between the hot equatorial regions and the cold polar regions drives the global circulation of air and sea. Warm air and water from lower latitudes tend to flow poleward, while cold air and water from higher latitudes tend to flow toward the equator. The path of both air and water is deflected by the Earth's rotation, helping to create the complex patterns of wind and air circulation. Great convective cells of rising and falling air form in each hemisphere, while the oceans convey heat through a global "conveyor belt" of interconnected warm and cold currents. Each medium carries roughly half of the total heat flux poleward. Not surprisingly, supercomputer models of the atmosphere or the ocean are usually referred to as general circulation models (GCMs).

The patterns of air and water circulation are closely interdependent. The winds moving over the sea surface drive the ocean currents, while the heat flux carried by the ocean currents supports the temperature differences that drive the winds. The topography of the continents and ocean bottoms imposes a complex set of boundary conditions on the atmospheric and oceanic circulations. Other significant factors have their effects as well: the greater heat-storing capacity of the oceans moderates atmospheric temperatures near the coasts, while clouds and sea ice cool both systems by reflecting sunlight. Another important factor in coupling the two systems is the hydrological cycle: the ocean provides most of the water vapor in the atmosphere; water vapor may be returned to the ocean in the form of fresh water fluxes from rain and rivers, which alter the salinity of the ocean, or it may be stored as the fresh water of lakes, soil, and ice.

In spite of the interconnection between the ocean and the atmosphere, the behavior of each system can be computed independently for many purposes. Instead of computing the realistic interactions, scientists insert a set of boundary conditions describing the air-sea interface.

Most of the contemporary GCM ocean models trace their roots to the pioneering work of Kirk Bryan and Michael Cox in the 1960s at the Geophysical Fluid Dynamics Laboratory in Princeton, New Jersey, administered by the National Oceanic and Atmospheric Administration of the Commerce Department. They formulated a finite-differenced model in spherical coordinates that included realistic coastlines and bottom topography. In the early 1970s, Albert Semtner at the National Center for Atmospheric Research (NCAR) dramatically restructured the Bryan-Cox model to take advantage of the supercomputers available at that time. The model was further refined in the 1980s by Semtner and Robert Chervin, also at NCAR, so that it could be run on any number of processors, thus making it easily adaptable to future generations of supercomputers. The restructured model, called POCM for Parallel Ocean Climate Model, ran at a sustained rate of 450 megaflops on NCAR's Cray X-MP and at a rate of at least 1.1 gigaflops on its eight-processor Cray Y-MP.

In the late 1980s, Semtner and Chervin used the Parallel Ocean Climate Model on NCAR's Cray X-MP to simulate the major currents of the world ocean. The model computes the temperature and velocity changes in the world's oceans in response to boundary conditions imposed at the sea surface. These conditions include the local wind stress, the seasonal temperature change in the atmosphere, and the influx of fresh water from rivers. Water temperature decreases most quickly with depth in the upper part of the ocean, and so to resolve vertical layers of differing temperature, Semtner and Chervin created a grid with uneven vertical spacing. The upper 10 levels extend to a depth of only 700 meters, while the remaining 10 levels cover a depth of about 5 kilometers. The highest 4 levels are only 25 meters thick. The simplified geography of the model ignores such details as the Mediterranean and attaches Madagascar to Africa.

A sensitive measure used to show the location of coherent ocean motions is the local temperature variability around a long-term average. Ocean currents and eddies have a high tempera-

ture variability because the bulk mixing induced by a current or eddy also mixes the temperatures of the water parcels. The overall global pattern of the currents is largely set by the average annual temperatures across the sea surface, which reflect the difference in solar heating between the equator and the poles. However, the seasonal cycle causes oscillations in the atmospheric temperature and wind direction that greatly amplify the ocean's temperature variability, especially in the tropics and in the Indian Ocean.

To separate the effects of these two different forcing factors, Semtner and Chervin have their model evolve the ocean for some time in response to only the average annual temperatures, then have their model evolve the ocean some more after adding the seasonal forcing. The local temperature variability around the five-year average at a depth of 37.5 meters, taking both effects into account, is illustrated on this page. High temperature variability around the mean, during both the annual and seasonal cycles, is found to be associated with such features as the Gulf Stream, the Kuroshio Current near Japan, and the East Australia Current.

Significant temperature variability is also seen where the Antarctic Circumpolar Current interacts with the topology of the ocean floor. Regions of high temperature variability are cre-

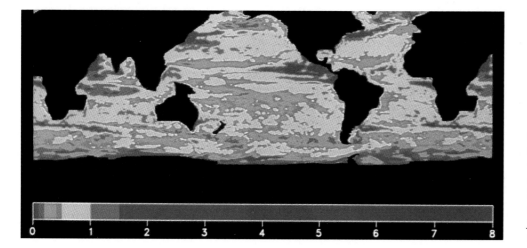

The major ocean currents stand out in orange and red in this image showing the variation in the computed ocean temperature. The color bar indicates the variation in degrees Celsius around a five-year average at 37.5 meters depth.

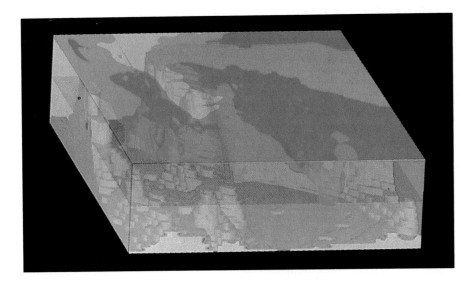

Variations in temperature around a five-year mean are caused by turbulent mixing as the Brazil and Falkland currents meet over rough bottom topography near the southern tip of South America. The dotted plane separates a magnified view of the upper 1 kilometer of ocean depth from the lower 4 kilometers. The pink surfaces surround volumes in which the variation is greater than 1°C, and the blue surfaces surround volumes in which the variation is only $\frac{1}{2}$°C.

ated by the confluence of the Brazil and Falkland currents in the southwest Atlantic and by the injection of warm water from the Indian Ocean into the South Atlantic off the tip of Africa. These effects are best observed by examining the temperature variability in three dimensions as shown above.

Warm currents impinging on cooler water generate eddies that are crucial for mixing. Eddies having warm cores break off from the unstable meanderings of the ocean currents to become isolated pockets of warm temperature. Because these warm pockets persist for a few weeks and are often rich in sea life, they are of great interest to the fishing industry.

It is estimated that the ocean eddies contain more than 99 percent of the kinetic energy of ocean movement. Yet, these analogs of the familiar atmospheric high- and low-pressure cells are typically 100 times smaller in diameter. Furthermore, they become smaller as one moves poleward. The finest spatial resolution feasible with NCAR's Cray X-MP, for instance, was $\frac{1}{2}$ degree in both longitude and latitude and 20 levels in depth, just barely fine enough to resolve the largest eddies in the equatorial regions.

To make further progress, scientists can narrow their studies to regional oceans and run the program for longer periods. Bill Holland and Frank Bryan at NCAR have modeled the Atlantic Ocean with a resolution of $\frac{1}{6}$ degree. This resolution gives their model nine zones for every one used by the Chervin-Semtner world ocean model. In addition, their model has 30 vertical levels instead of 20. The higher resolution, shown on the next page, not only brings features such as the Gulf Stream into better focus but, more important, begins to resolve the larger eddies. The model is approaching the true complexity of the Gulf Stream as revealed in the satellite image on page 168 of the temperature distribution throughout the northwestern Atlantic Ocean.

Predicting the Weather

Reliable weather forecasting has long been the dream of meteorologists. The first person to attempt that goal with a computer was the British scientist Lewis Fry Richardson. While serving as an ambulance driver during World War I, Rich-

A map of the December sea temperature of the Atlantic Ocean at a depth of 18 meters reveals the formation of eddies by the Gulf Stream. Water temperature decreases from red to blue. Compare this ocean simulation of high spatial resolution with the temperatures recorded by satellite, shown on page 168.

ardson spent time between battles trying to forecast the weather using a mechanical calculator. The machine's inadequacies were soon obvious and, in 1922, Richardson published the now-classic book *Weather Prediction by Numerical Processes* in which he described a human-based supercomputer that might have the required power.

Richardson's computing factory consisted of 64,000 people seated in galleries in a huge spherical room. A map of the Earth painted on the walls defined the computational mesh with a spatial resolution of one degree. Each grid zone was assigned to a single person, who calculated the state of the atmosphere at his or her own grid point using data provided by the nearest neighbors. A conductor, much like that of an orchestra, was responsible for synchronizing the calculations. Instead of waving a baton, the con-

ductor would shine a rosy beam of light on those people who were ahead of the rest, and a blue light on those who were behind. Richardson's vision foreshadowed the advent of massively parallel supercomputers, which are today rising to the challenge of climate modeling. Nevertheless, he severely underestimated the magnitude of the computing power that is required to reliably predict the weather.

Supercomputers with teraflop processing power are generally believed to be the machines that will fulfill Richardson's dream. Computing at teraflops speed (10^{12} floating point operations per second) for three hours (about 10^4 seconds), a machine performs a total of about $10^{12} \times 10^4 = 10^{16}$ floating point calculations. To place this colossal amount of arithmetic in perspective, consider that a human being needs a few minutes (about 10^2 seconds) to perform one floating

Human computers seated in galleries around a conductor diligently perform the computations needed to predict the weather. The results of these calculations are lowered in a bucket to a telegrapher, who then transmits the forecasts to cities around the world.

point operation: the addition or multiplication of two 13-digit numbers. The computing power of a human being is therefore about 10^{-2} flops. To perform 10^{16} calculations, a single person needs $10^{16} \div 10^{-2} = 10^{18}$ seconds, which is about 20 billion years, the approximate age of the universe. No wonder that scientists covet the unprecedented speed and precision of massively parallel supercomputers.

To predict the weather using a modern atmospheric general circulation model, meteorologists must begin with detailed information about the atmosphere at some instant. These data constitute the initial conditions that are the starting point from which a supercomputer calculates forward in time to simulate the weather's future evolution. Values for the initial conditions come from observations from a variety of sources, such as balloons, ships, airplanes, and Earth-orbiting satellites.

All these data are distributed by a global telecommunications network to major weather forecasting centers such as the Canadian Meteorological Centre, the United States Air Force

Global Weather Central, and the European Centre for Medium-Range Weather Forecasts (ECMWF) in Britain. There the results of the previous prediction of the global state of the atmosphere are blended together with incoming observational data to construct initial conditions for the next analysis. The supercomputer-generated prediction of temperature, pressure, wind velocity, and water content at each point in space smoothly fills in the large regions for which no measurements are available. The blending of supercomputer predictions and observational data, called data assimilation, produces a set of initial conditions that is more reliable and self-consistent than that which could be gathered from purely observational data alone.

The ECMWF prediction code uses a spectral model that resolves the atmosphere horizontally into waves rather than grid points. Every circle of latitude is surrounded by 106 waves, corresponding to a grid resolution of 1.1 degrees—almost exactly the resolution envisioned by Richardson, seventy years ago. Vertically, the atmosphere is divided into 19 levels. Unlike the ocean, which takes many years to mix, the atmosphere mixes in only hours, as we saw in the thunderstorm example in Chapter 1. Thus, a lot of effort goes into accurately modeling the physical processes that drive the vertical convective motions of air parcels in the lower atmosphere. At the ECMWF, weather forecasting is carried out in parallel on an eight-processor Cray Y-MP, at sustained speeds in excess of a gigaflop.

A global weather forecast prepared by the ECMWF provides an example of the program in use. Starting with the weather at noon on Sunday July 28, 1991, the program computed the evolution of the global wind circulation for three days into the future. The map for the final day, July 31, shows the predicted global arrangement of wind patterns and temperatures. Because air flows from high-pressure regions toward low-pressure regions in order to equalize air pressure, the direction of air flow reveals multiple high- and low-pressure centers around the globe. Since the Earth is rotating, the moving air expe-

riences a force, called the Coriolis force, that deflects the air flow from a straight path. As a result, winds in the Northern Hemisphere blowing toward a low-pressure region rotate counterclockwise about the low, forming a cyclone. Winds blowing away from a high-pressure region rotate clockwise about the high, forming an anticyclone. In the Southern Hemisphere, the directions are reversed.

Although the model captures the broad features of the atmospheric circulation, it does not give very detailed predictions for smaller regions such as the various countries of Europe. To attain better local forecasts, meteorologists use the global forecast as the boundary condition for a shorter-range regional forecast having a higher spatial resolution. The maps on page 180 show the results of a regional weather simulation covering Europe, the Mediterranean, and most of the North Atlantic, also produced on the Cray Y-MP at ECMWF using their operational model.

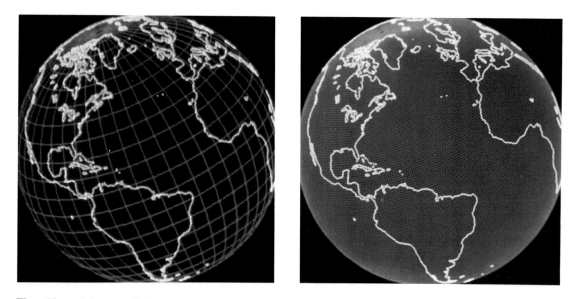

The grids used in atmospheric models today have resolutions that range from 8 degrees (890 kilometers at the equator, left) down to 1 degrees (120 kilometers at the equator, right). The coarser grids are typically used in climate codes and the finer ones in weather prediction.

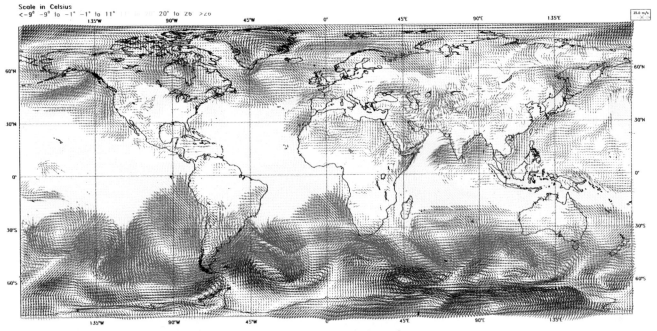

A three-day weather forecast produced this map of the predicted wind velocity and temperature at an elevation of 30 meters above the surface of the land or sea. The length and orientation of an arrow indicate wind speed and direction. Its color represents air temperature, ranging from violet for less than −9°C to red for more than 26°C.

The weather at noon on Sunday, July 28, 1991, again serves as the initial conditions. Twenty-four hours later we can see how the wind velocity and temperature have evolved. During that time, a low-pressure cell (shown with yellow wind vectors) in the center of the map has strengthened and caused rainfall as it moves toward the British Isles.

The results of such computations reach the public as regional, aviation, and marine forecasts. As the speed of supercomputers increases and the algorithms are adapted for vector and parallel computers, the resolution of the models becomes higher. The accuracy of the weather forecasts steadily improves, not only because of higher resolution grids, but also because of better analysis procedures, more realistic physics in the programs, and better collection of data for the initial conditions. For instance, the 1987 ECMWF five-to-six-day forecast was more accurate than a three-day forecast made on 1977 supercomputers. We can expect accuracy to continue improving as the explosive growth of supercomputers proceeds in the 1990s.

Climates of the Future

There is a fundamental limit to the period of time for which one can hope to make an accurate weather forecast. As MIT scientist E. N. Lorenz pointed out in 1963, the equations governing convection give rise to chaotic solutions. Small changes in initial conditions cause solutions to diverge with time, or more poetically put, "a butterfly flapping its wings in California can eventually influence the weather in New

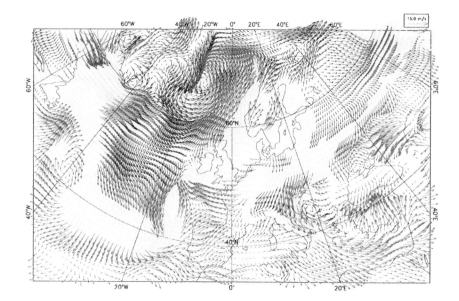

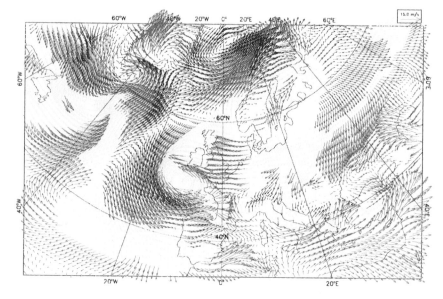

The map at the bottom shows the predicted evolution of the weather over Europe and surrounding regions during the 24 hours after measurements were taken for the map on top. The maps show wind velocity and temperature at 30 meters elevation.

York." Because atmospheric convection plays such a large role in the weather, it is thought that accurate weather predictions are not possible for time periods exceeding two weeks into the future.

Fortunately, climate models do not share this problem, since they do not attempt to predict individual weather events, but instead only statistical averages such as temperature means or seasonal fluctuations. Thus, supercomputer programs combining atmospheric and ocean models can, in principle, simulate climate changes over periods ranging from years to hundreds of millions of years.

This supercomputing discipline gradually emerged over many decades from its beginnings in the weather prediction codes pioneered by Richardson. Many climate modeling groups exist today throughout the world, working with a number of major supercomputer centers whose major focus is climate research. We will be able to profile only a few researchers' results, but we hope to illustrate the complexity of the interactions in the Earth's climate system. One major goal of climate research is to predict through simulations the global rise in temperature that may result from the enhanced greenhouse effect. Greenhouse warming is caused by gases that are transparent to visible light, but absorb infrared radiation. Sunlight easily penetrates the atmosphere and warms the ground, but the infrared radiation emitted by the warmed Earth is trapped by these greenhouse gases. Water vapor accounts for about 80 percent of the natural greenhouse effect. Indeed, water vapor has been creating a greenhouse effect for most of the history of life on the planet; without its presence, much of the Earth would be frozen. Scientists and others concerned about the greenhouse effect are not worried by this natural and beneficial component of the atmosphere, but rather by the rapid increase in other greenhouse gases, such as carbon dioxide (CO_2). The presence in the atmosphere of CO_2 is increasing through the burning of fossil fuel and other human activity.

Unfortunately for the shapers of public policy, the buildup of greenhouse gases in the atmosphere is not the result of a single human activity. The National Academy of Sciences has recently released a report finding that the use of fossil fuel accounts for less than half the increase in the greenhouse effect. One group of contributors are the chlorofluorocarbons (CFCs) used as refrigerants, which also deplete stratospheric ozone. Their replacement over the next few years has been mandated by the Montreal Protocols. Another contributor is methane released from cattle, rice paddies, and nitrogen fertilizer. Finally, the burning of forests to clear land for farming not only adds carbon dioxide to the air, but also removes living sinks of atmospheric carbon dioxide.

Of all these gases, CO_2 has the largest impact on global warming, because of the immense volume of the gas now being released. Before the start of the industrial age two hundred years ago, the concentration of CO_2 in the atmosphere was about 280 parts per million. As a result of industrialization, deforestation, and new forms of transportation, CO_2 concentration has risen by 25 percent to 355 parts per million today. If present trends continue, the atmosphere in the middle of the next century will have at least doubled the CO_2 content of the preindustrial atmosphere of the 1800s.

To examine the environmental consequences of the continuing increase in greenhouse gases, many research groups have used general circulation models of the atmosphere and the oceans. Most of these climate models still run on vector multiprocessor supercomputers, often in parallel mode, but the United States Department of Energy has an aggressive program called CHAMMP to create a new generation of climate codes that run on massively parallel supercomputers. We shall examine several approaches to climate modeling to show how the physical complexity of the model can have a major impact on its predictions of global warming.

Warren M. Washington, director of NCAR's Climate and Global Dynamics Division, and his colleagues performed climate simulations in the mid and late 1980s, using a modified atmospheric general circulation model. An intimidating number of factors influence the course of greenhouse warming. Washington's model was able to take into account many of them, including the incoming solar radiation and the changes in heat caused by precipitation and evaporation. It also considered the effects of clouds, which reflect sunlight before it reaches the Earth, and of sea ice and snow, which reflect sunlight rather than absorb it.

However, some crucial simplifications had to be made so that the model would run on the available supercomputer, a Cray-1 at the National Center for Atmospheric Research. The ocean was approximated as a mixed layer 50 meters thick; thus the model omitted any heat exchange with the deep ocean or any realistic current or eddy patterns. Both of these factors strongly affect the ocean's ability to slow global warming by absorbing heat. The atmosphere was divided into nine unequally spaced vertical levels extending from the ground up to an elevation of about 30 kilometers. The horizontal grid resolution was 4.5 degrees in latitude and 7.5 degrees in longitude, much cruder than the resolution for the weather forecast described above.

To examine the impact on world temperatures of a doubling in CO_2 concentration, Washington and his colleagues performed two simulations, one a "control simulation" specifying the current CO_2 concentration as an initial condition, and the other an "experimental simulation" specifying twice today's value. Aside from their CO_2 concentrations, both simulations started from the same near-equilibrium initial conditions. Both were evolved forward in time for a total of 19 years. The Cray-1, which was about to be retired, required nearly three weeks to run the two simulations.

The control simulation reproduced the seasonal changes and the large-scale atmospheric circulation patterns characteristic of today's climate. Its success gives confidence in the experimental simulation, and some basis for believing that the results of the doubled CO_2 simulation are broadly correct. The simulation finds that an instantaneous doubling of the carbon dioxide in the atmosphere has large effects. The average surface temperature of the Earth increases to a value 4.5°C warmer than today's average. The effects of a few-degree change in temperature become clearer when we consider that the world was about 4°C cooler on average in the midst of the last ice age.

A visualization of the simulation's results carried out in collaboration with Jeffery Yost of the National Center for Supercomputing Applications, showed that the amount of warming varies from region to region. In particular, the continents experience greater warming than the oceans. Some regions end up with temperatures that are cooler than normal, while others experience twice as much warming as the average. The greatest warming appears at high latitudes. As the Earth's average temperature rises, the snow and sea ice covering the polar regions

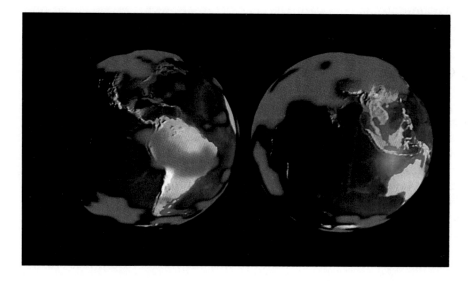

Washington's simulation predicts that in the regions shown in red, the surface air temperature will be more than 5°C warmer than today, twelve years after the instantaneous doubling of carbon dioxide in the atmosphere.

begin to melt faster than they are replaced. As the snow and ice recede, the reflectivity of the polar regions decreases. The polar ground can absorb more sunlight, and the rate of melting further accelerates.

As more powerful supercomputers became available in the late 1980s, Washington and other scientists were able to couple their atmospheric models to more realistic and computationally demanding ocean models. For example, a coupled ocean-atmosphere model was used by Ulrich Cubasch and his Hamburg colleagues at the Max Planck Institute for Meteorology, the Meteorological Institute of the University of Hamburg, and the German Climate Computing Center to reexamine the effect of deep oceans on the Earth's response to increasing concentrations of atmospheric carbon dioxide. They borrowed a modified version of the ECMWF weather prediction code, but like Washington, they used a much coarser zoning of 5.6 degrees. Because the ocean model would be too coarsely zoned to resolve eddies, the German group ran two different ocean models to bracket the uncertainties introduced by the coarse zoning. The atmospheric model provides the values of heat, fresh water, and wind stress that determine ocean currents and the exchange of water between the air and sea. The ocean model in turn provides the values of sea surface temperature and ice thickness that serve as boundary conditions to drive the atmospheric model.

Cubash's group performed several numerical experiments on the Hamburg Cray-2 to examine the long-term results of various policies that might be adopted to respond to global warming. One strategy, called Scenario A, is the "business-as-usual" scenario prepared in 1990 by the Intergovernmental Panel on Climate Change (IPCC). The concentration of CO_2 in the atmosphere continues to rise gradually at the current rate, from an initial value of about 350 parts per million in 1985. The concentration doubles by 2020 and continues to grow until it reaches nearly 1200 parts per million in 2085 (quadruple the preindustrial value!).

Because the deeper ocean of Cubasch's model can store some of the heat being generated, we expect the temperature to rise more slowly than the simpler models predict. Indeed, when the Hamburg group computed the results of instantaneously doubling the CO_2 concentration in a preliminary experiment similar to Washington's, they found that in thirty years the climate warmed between 1.5 and 2.6°C, depending on the ocean model. This value is considerably lower than Washington's and testifies to the ocean's great influence on climate.

Performing another simulation of gradual CO_2 buildup, the Hamburg group found that Scenario A would lead to an average global warming of 2.7°C during the next 100 years. The region of most pronounced temperature rise (greater than 6°C) lies within the Arctic Circle, as also found in Washington's simulation. The Great Plains and central Canada become warmer, as do desert regions such as the Sahara and central Australia. The sea ice cover and the cloud cover are slightly reduced. More precipitation falls over land.

The expansion of the oceans as they warm produces a rise in sea level of about 16 centimeters, although again sea level rises more slowly than the simple ocean model predicts because the thermal inertia of the deep ocean prevents it from heating quickly. In some locations, the rise in sea level may exceed this average value by a factor of two. Clearly the slow absorption of heat by the deep ocean is crucial to determining the course of atmospheric warming. The Hamburg group's simulation shows a thermal front of higher temperature advancing downward into the Pacific Ocean. However, even a century from now, this front will not have penetrated the broad volume of the deep ocean. The climate will still be evolving in response to today's human activities for many generations.

The recent Earth Summit on the environment in Brazil focused the attention of the world on the need to rethink our activities. To explore the alternatives that face humanity, Cubasch and his colleagues ran a simulation,

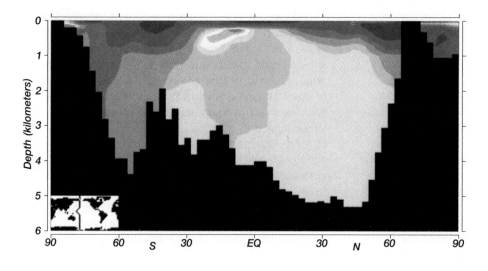

The largest temperature increases occur within 1 kilometer of the ocean's surface. This vertical slice through the Pacific Ocean shows the annual mean temperature changes during the final ten years of the Scenario A simulation, from 2076 to 2085. The black regions indicate the profile of the ocean floor, extending from the south pole at the left to the north pole at the right. Colors indicate temperature change, with dark brown to pale gold indicating temperature increases and yellow and blues representing temperature decreases.

called Scenario D, that assumes that the governments, citizens, and corporations of the world immediately take draconian measures to reduce the release of CO_2 into the atmosphere. By replacing fossil fuels with renewable energy resources, the world's nations stabilize the concentration of greenhouse gases by 2025. The CO_2 concentration levels off at about 560 parts per million by 2085. At the end of the simulation, the atmosphere has about half the CO_2 of the atmosphere predicted by a business-as-usual scenario. If we take the extraordinary measures demanded by the optimistic Scenario D, the average global temperature increases by only about 0.75°C in 2085 and the sea levels rise by about 5 centimeters.

Even this lower rate of global temperature rise is enormously high by historical standards. The 4°C rise in global temperature after the end of the last ice age took some ten thousand years, a rate one hundred times slower than the rate predicted by Scenario D. Not all regions of the biosphere may have time to adapt to this rapid warming. An ecosystem requires many generations of reseeding to move northward. When temperature and precipitation change in less than one hundred years, many species will die out simply because they cannot move their

econiches fast enough. To understand this larger problem, we will require climate codes coupled to models of regional ecologies, a topic addressed at the end of the chapter.

It is a matter of considerable scientific controversy whether the present generation of coupled ocean-atmosphere codes are reliable enough to justify drastic immediate action, at enormous expense, to limit the world's CO_2 production. Numerous uncertainties could affect the predictions of these codes. How will algorithms that represent cloud cover in much greater detail affect the model's outcome? Could effects that are small on average over a global scale nonetheless have large consequences for a particular region? Could some factor not now included in the models set off abrupt changes rather than the slow changes so far predicted? Would the reduction of coal burning reduce not only CO_2, but also sulfur dioxide aerosols, which act as cooling agents?

Finally, there is considerable disagreement about how to interpret the observational data. Does the fact that the warmest eight years in the last century all fell in the last twelve years signify that greenhouse-induced warming has been detected? How does one separate the short-term cooling caused by occasional volcanic injection of aerosols into the stratosphere, such as we are now experiencing from Mt. Pinatubo, from longer warming trends?

In 1992, the Intergovernmental Panel on Climate Change reviewed climate researchers' models, taking into account a number of these uncertainties. In particular, the panel considered the cooling effects of both stratospheric ozone depletion and the increased emission of sulfate aerosols; the consequences of reducing chloro-fluorocarbon emission; the lag effect of oceanic thermal inertia; and the difference in emissions between the Northern and Southern hemispheres. The panel also considered the enhanced plant growth caused by higher levels of CO_2; by absorbing carbon from the atmosphere, the additional plant life could reduce the concentrations of some greenhouse gases. Most of these effects reduce global warming and the expected sea level rise from their levels in the panel's 1990 study.

The Intergovernmental Panel on Climate Change does not use a full-scale global climate model to predict the amount of warming. Rather, it relies on a very simplified set of equations that captures the broad effects of the many physical processes explored in the global climate models. The results of their new predictions for the next century lie between Cubasch's results for scenarios A and D, in spite of the new effects included. The same model now provides an approximate fit to the global temperature averages of the last century.

A debate is now raging to resolve how drastically the world should pursue the reduction of greenhouse emissions. Michael Schlesinger, a GCM climate modeler at the University of Illinois at Urbana-Champaign, and his colleagues have recently argued for a two-phase approach. Since the cost of rapidly reducing CO_2 emission throughout the world could run into the trillions of dollars, Schlesinger argues that it would be a wise investment to develop by the mid-1990s a new generation of teraflop climate codes with an unprecedented horizontal spatial resolution of 50 kilometers. These codes

would be required to reduce the uncertainty in estimating the costs of climate change. During this time, less-costly measures such as energy conservation should be adopted, while the decision to switch to new fuels or pursue other aggressive means of reducing emissions would be made only after more-accurate cost/benefit computations show them to be warranted.

In spite of all the debate, most participants seem to agree on the need to accelerate efforts to refine supercomputer models of climate change, while taking limited precautionary measures to curtail the burden our growing population is placing on the Earth's atmosphere, oceans, and biosphere.

Climates of the Past

A supercomputer can simulate ancient climates as well as future ones. When a simulation models past climates already known from chemical, geological, and biological evidence, it serves as a test of the veracity of climate codes. A number of groups are currently engaged in this type of research project. These groups are exploring how Greenland became ice-free during the Viking explorations 1000 years ago, why there have been periodic ice ages during the last 100,000 years, and whether climate was a primary agent in global mass extinctions millions of years ago.

John Kutzbach and Robert Gallimore at the University of Wisconsin Center for Climatic Research have simulated climate conditions 250 million years ago, using a low-resolution general circulation model on the Cray X-MP at the National Center for Atmospheric Research. This important moment in the Earth's history fell at the close of one great geobiological era, the Paleozoic, and the start of another, the Mesozoic. At the exact transition point 248 million years ago, the greatest mass extinction in the history of multicellular life destroyed more than 95 percent of the biomass on Earth.

Mammal-like reptiles called therapsids were the dominant land animal during the Permian Period, the last period of the Paleozoic Era before the great extinction. The therapsids are believed to have been warm blooded, and perhaps they thrived because of their consequent ability to adapt to large temperature extremes. Had the great extinction not wiped out most of their genera, the Age of Mammals might have begun two hundred million years earlier than it did. After the extinction, other reptiles rapidly evolved into dinosaurs, which dominated the land until another mass extinction at the end of the Cretaceous Period 65 million years ago. The few remaining therapsid genera eventually evolved into mammals, finally beginning their reign after the extinction of the dinosaurs.

The cause of the mass extinction at the end of the Paleozoic remains a mystery, but one of the leading candidates is climate change induced by continental drift. By the end of the Permian Period, all of the major continents had drifted together into one huge landmass called Pangaea, centered on the equator but spanning one side of the globe from pole to pole. The closing of many sea lanes drastically altered ocean currents and thus the effects of the currents on the atmosphere. Many continental shelves disappeared as the continents joined, and thus many of the primary sites of ecological diversity in the oceans were lost. The patterns of temperature and rainfall over the huge landmass changed considerably. Later, Pangaea split into two supercontinents, Laurasia and Gondwanaland. Laurasia eventually broke apart to form North America, Greenland, and Eurasia, while Gondwanaland formed South America, Africa, India, Australia, and Antarctica.

Kutzbach and Gallimore's climate model shared many of the simplifications of Washington's model. In addition, the computational grid was about twice as coarse in each dimension as Washington's. As was appropriate to the relatively low-resolution grid, Kutzbach and Gallimore adopted a simplified geometry for the Pangaean continent. Pangaea was assumed to be symmetrical about the equator, and its west coast was oriented exactly north-south. In spite of these deficiencies, trial simulations of today's climate and landmasses gave results in broad agreement with observations. Kutzbach and Gallimore could be optimistic that the simulated Pangaean climate would also capture the main features of this bygone period.

After summarizing the results of their simulation as seasonal averages, they found that the temperature of Pangaea's interior showed more extreme variation than the temperature of today's continents. Because Pangaea consisted of one enormous landmass, a large percentage of the continent was far removed from the moderating effects of the ocean. The interior therefore became hot in the summer as the average temperature climbed to 35°C (95°F), and cold in the winter when the average temperature fell to −38°C (−36°F). In contrast, the great ocean Panthalassa had surface temperatures and sea-ice distributions similar to those of modern oceans.

The summer heat drove vast, rotating low-pressure air cells that pulled moist ocean air over the land, creating intense monsoons. An image on the facing page from the simulation shows a summer monsoon low centered just west of the region of maximum summer heating near 35°N. By calculating precipitation and subtracting evaporation, Kutzbach and Gallimore showed that most of Pangaea's interior was arid or semiarid. A reasonable speculation is that the shift to this harsh climate may have driven the therapsids and other land animals to extinction by altering the plant communities on which they depended.

The complex codes of the next few decades will couple much more refined climate models to the mantle dynamics codes described at the beginning of the chapter, allowing scientists to explore the impact of continental drift on world and regional climates. In addition, these codes will be able to explore events that take place on much shorter time scales, such as collisions of

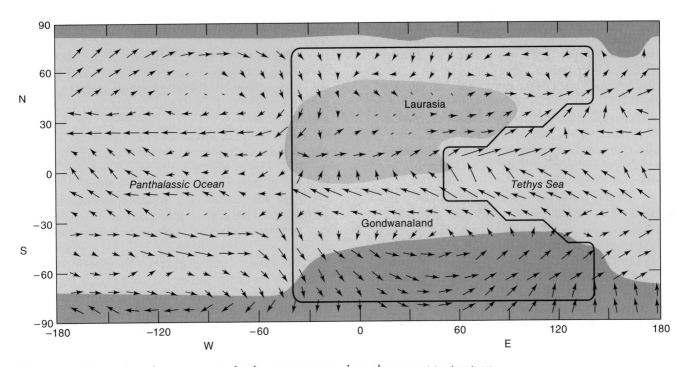

This map of Pangaea's surface temperature for the summer season shows the warmest temperatures (over 30°C, or 86°F) occurring in mid-summer over western Pangaea inside the yellow/orange region. Over the ocean, the 0°C contour (blue regions) approximately locates the extent of sea ice. The wind flow patterns indicated by the arrows bring increased tropical rainfall during the summer months. The illustration shows the simplified geometry used by Kutzbach and Gallimore to represent the supercontinent. Today's Mediterranean Sea is a surviving remnant of the ancient Tethys Sea.

comets or meteors with the Earth. Detailed models of the drastic climate alteration triggered by such events may show why, in spite of vast extinctions, enough of Earth's life forms survive to reestablish a flourishing global ecology.

The Chernobyl Accident

Whereas climate changes so slowly that we may be tempted to ignore its effects, scientists also model events that happen so fast our attention is riveted. The reactor meltdown in Chernobyl, Ukraine, six years ago is a good example. Shortly after 1:00 a.m. on the morning of April 26, 1986, an explosion of hydrogen gas ripped apart a nuclear reactor at the Chernobyl power station about 70 miles north of Kiev. The explosion blew the roof off the reactor building and lofted radioactive debris—including extremely radioactive fuel—into the atmosphere.

Between 30 and 50 megacuries of the radioisotope iodine-131 was dispersed into the environment. In contrast, the Three Mile Island accident in the United States released only 0.00001 of a megacurie of iodine-131, and the detonation of a 20-kiloton nuclear weapon releases about 2 megacuries. Iodine-131 is particularly troublesome to human beings because it is readily absorbed by the thyroid gland. After the crisis was over, Janusz Pudykiewicz of the Atmospheric Environmental Service of Canada used a supercomputer model to study the atmospheric transport of three radioactive isotopes, iodine-131, cesium-137, and xenon-133. The same supercomputer helped him compare the program's predictions with the measurements made during the emergency.

To drive the simulation, Pudykiewicz inserted data describing atmospheric circulation patterns throughout the Northern Hemisphere, computed on a Cray X-MP supercomputer at the Canadian Meteorological Centre. Numerical analyses of weather data performed every few hours at the center provided a remarkably complete picture of the meteorological field, including wind velocities and water content at various elevations in the atmosphere. Pudykiewicz's

computer code is termed a tracer program because the radioactive particles trace out the air flows in the atmosphere, but do not alter the air flows by their presence. Thus the radioactivity may be ignored in computing the evolution of the atmosphere, unlike the case of global warming in which the CO_2 buildup actually alters the climate state.

Pudykiewicz's tracer code includes a mass conservation equation to keep track of the transport of radioactive debris, and it considers the rate of radioactive decay of one isotope into another. The code takes into account the rate at which new radioactive material was introduced into the atmosphere by continuing fires at the power plant, and the rate at which radioactive material left the atmosphere, either because particles fell to Earth under the influence of gravity (sedimentation) or were cleansed from the atmosphere by rain and snow (scavenging).

As can be seen on this page, the resulting simulation demonstrated that the cloud moved

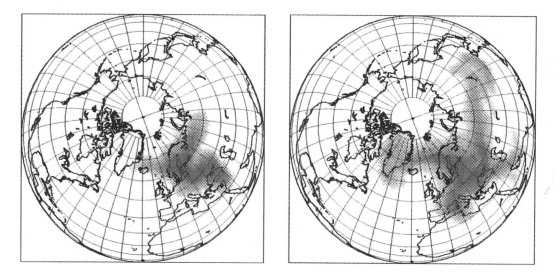

These two images show the rapid transport of the radioactive cloud from the Chernobyl accident. On April 28, the cloud crossed Scandinavia and entered a cyclonic flow that propelled the debris toward Greenland. Meanwhile, a high-pressure system of Russia dissipated, allowing the cloud to expand eastward. By April 30 the cloud stretched across Asia and had just reached the northeastern coast of North America.

both eastward and westward following the air flows during the period after the explosion. To verify the simulation's accuracy, Pudykiewicz compared the predicted concentrations of radioisotopes with actual concentrations measured at 17 stations around the world, including Stockholm, Helsinki, Paris, Halifax, Vancouver, and Winnipeg. The model accurately predicted the time that radioactive debris arrived at the various sites, and, in general, it duplicated the variations in radioactivity subsequently measured at these sites. For example, the model accurately tracked the rapid decrease in the intensity of the radioactivity in Stockholm by a factor of 100,000 over the first two weeks.

Pudykiewicz's simulation, as well as those at other supercomputer centers, demonstrates that three-dimensional tracer models can indeed be important tools during future global emergencies, such as nuclear or chemical plant accidents. These models are flexible enough either to analyze normal meteorological fields or to warn people of impending danger in an emergency.

Los Angeles Smog

Photochemical air pollution plagues cities and urban areas around the world. In the United States, for instance, more than half of the inhabitants breathe air whose ozone content exceeds limits set by the government, even though the nation spends over $100 million per day to improve air quality. Pollution is considerably worse in many cities throughout Eastern Europe, Asia, and Latin America.

Traditionally, the quality of air has been improved by controlling the emission of pollutants. For example, the phasing out of leaded gasoline has drastically reduced lead pollution, which can cause neurological disorders. Yet this strategy is not easily applied to the problem of air pollution, because the chemical compounds, such as ozone, that pollute the typical urban atmosphere are not directly emitted but are pro-

duced by the action of sunlight on nitrogen oxides and reactive hydrocarbons, which come from high-temperature fuel combustion, solvents, and automobiles.

In the 1980s, Gregory J. McRae and Armistead G. Russell at Carnegie Mellon University began using supercomputers at the Pittsburgh Supercomputing Center to explore the chemistry of ozone formation over the Los Angeles basis, home to more than ten million people. Boundary and initial conditions applied to the problem include the topography of the Los Angeles area and the locations of emission sources, including both fixed points such as power plants and geographically distributed sources, such as the freeway system. The boundary conditions even included information on how emissions from these sources varied over time. The model also factored in the changes in solar illumination and wind-flow patterns that take place over a 24-hour day.

The computational model developed by McRae and Russell has a horizontal resolution of 5 kilometers and a variable vertical resolution, creating a mesh of about 12,000 grid points. More than 500,000 equations were necessary to characterize the chemical evolution of 50 compounds involved in pollution at each grid point in the air over Los Angeles. Even though the code runs in parallel on the Cray Y-MP at a speed of more than a gigaflop, McRae and Russell use several hundred hours a year on this problem.

In a series of more than 45 simulations that assumed various ratios of nitrogen oxide and hydrocarbon emissions, McRae and Russell demonstrated that the regulatory emphasis on the control of reactive hydrocarbon emissions alone is inadequate. The atmospheric chemistry is so complicated that the simple-minded approach of reducing only one emission does not necessarily reduce the amount of ozone in the atmosphere. Also, as vividly portrayed in a animation created by NCSA graphics specialists Michael McNeill, William Sherman, and Mark Bajuk, pollution produced in one location can

This visualization of McRae and Russell's results shows that air quality is reduced far downwind of the polluting sources. Nitrous oxides and hydrocarbons emitted in the vicinity of downtown Los Angeles are transported inland by prevailing off-shore winds (blue arrows). Sunlight drives a host of chemical reactions that produce the ozone that heavily pollutes the air of western San Bernardino County, about 25 miles east of Los Angeles. Each green particle represents 10 tons of reactive hydrocarbon gases, each yellow particle 10 tons of carbon monoxide, and each red particle 10 tons of nitrogen oxides.

be carried by the wind to the detriment of air quality in neighboring locations. In order to improve air quality significantly, the emission of both hydrocarbons and nitrogen oxides must be controlled throughout the Los Angeles area.

McRae and Russell's code can evaluate various costly strategies to learn whether they would actually alleviate the problem. For instance, one such simulation demonstrates that the conversion to vehicles fueled by methanol could bring the entire Los Angeles basin close to compliance with current standards. Not only the ozone, but the emission of fine-gained particulates would be reduced dramatically. These particles, which are responsible for degrading the transparency of the atmosphere, are suspected of carrying carcinogens. Similar computa-

tions are being carried out by McRae, in collaboration with researchers in Mexico, to simulate air pollution in Mexico City. Ultimately, cities throughout the world will have this new tool to help them make decisions crucial to creating a livable environment.

Human Interactions with Ecological Systems

Ecologists are beginning to create computer programs that can be used to study the interrelationships of living organisms with complex environments. These programs have an immediate practical value because they can be directed toward the study of alternative policies of human intervention. Although many interventions, such as the suppression of forest fires or the control of river flooding, are seemingly benign, every such intervention upsets a carefully balanced web of interactions that has built up over thousands of years. We shall study two attempts to model such interactions, one involving forests and the other wetlands.

The devastating 1988 forest fires in Yellowstone National Park set off a national debate on fire suppression policy. For the previous one hundred years, the United States government had followed a program of extinguishing all forest fires, whether started by human or natural agents. Many have assumed that this is clearly the correct policy. However, fire suppression leads to the dramatic loss of landscape diversity, a consequence illustrated by research carried out by David Kovacic, a professor at the University of Illinois at Urbana-Champaign; William Romme, a professor at Ft. Lewis College in Colorado; and Don Despain, an ecologist at Yellowstone National Park.

The scientists obtained a three-hundred-year history of the effects of fire on the plant life of Yellowstone from nondestructive tree core samples taken from a study area composing about 20 percent of the land area of Yellowstone. During the first two hundred years, all fires had been allowed to rage freely. The resulting series of small fires created patches of diverse age that offered animals a wide range of habitats. After 1872, however, a policy was introduced to suppress all fires. By 1988, this policy had created a large area of old forest with a floor covered with a large mass of flammable material. Reduced landscape diversity had established favorable conditions for large fires to start at the onset of high winds and drought.

Working with Allan Craig, an NCSA visualization staff member, Kovacic and his colleagues first created an animation showing the actual evolution of the landscape over three centuries, as determined by the tree core samples. The researchers then ran a simulation on NCSA's Cray X-MP supercomputer showing the consequences to ecological diversity if all fires had been suppressed during this historical period. They found that the suppression of all fires creates an abnormal situation in which the landscape is of uniform old age. There is less habitat diversity for plants and animals, and the accumulation of flammable material sets the stage for very large fires that may be impossible

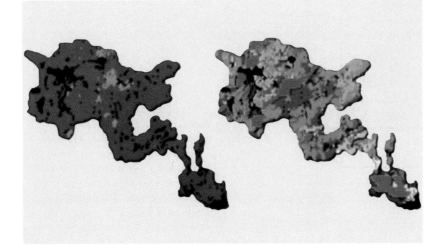

A comparison of the present diverse ecology of Yellowstone (right) to a simulation showing the result of suppressing fires for the last three hundred years (left) demonstrates that the absence of fire creates a forest of single-age growth. The age in years of the trees is indicated by color coding: yellow (0–40), light green (40–150), medium green (150–300), and dark green (300+). The regions shown in red represent the regions burned by the 1988 fires.

to contain. Kovacic's new computational tools will aid in developing a new set of policies that take into account the vital role of fire in the ecology of forests.

Models of coastal wetlands are quite complex because they must factor in hydrology, chemistry, and physical topography. A good example is the effort of Robert Costanza of the University of Maryland, Fred Sklar of the University of South Carolina, and Mary White of Louisiana State University to model a portion of the Mississippi River delta marshes called the Atchafalaya River basin.

The Atchafalaya delta and the adjacent Terrebonne Parish marshes represent one of the most rapidly changing landscapes in the world. The United States Army Corp of Engineers has for many decades built dams and levees to control the flow through the Mississippi River and its tributaries. For the last forty years, the Corp has allowed about 30 percent of the Mississippi River flow to move down the Atchafalaya, which continues to nourish the wetlands. However, the flow of sediment-laden water into the rest of the Louisiana coastal marshes has been greatly curtailed. As a result, a large area of wetlands has been lost to sediment starvation and salt water intrusion. The Corp is now considering extending a levee along the east bank of the Atchafalaya, which would restrict the flow of water and sediment into the Terrebonne marshes.

The Coastal Ecosystem Landscape Spatial Simulation (CELSS) model was developed by

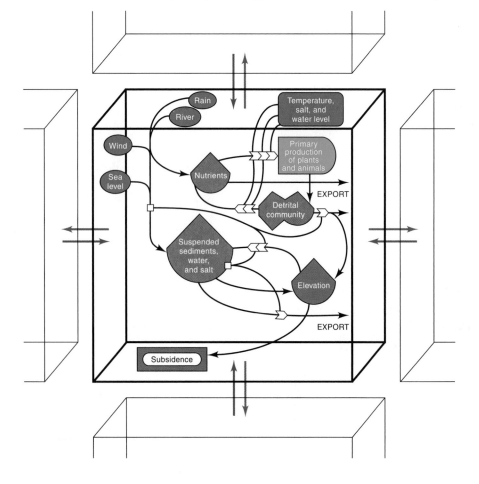

The wetland area is divided into cells that exchange flows containing suspended sediments and chemicals with their nearest neighbors. In each cell a set of coupled equations describe biological, hydrological, chemical, and physical processes. The arrows schematically show the interrelationship between the key variables.

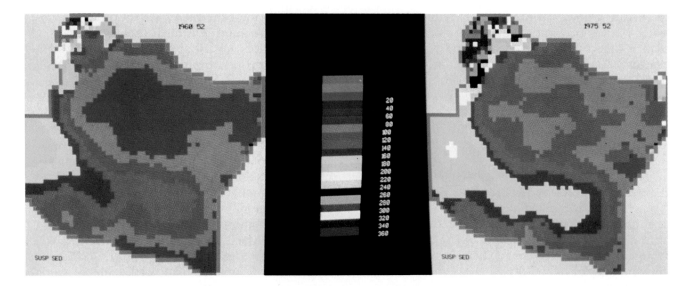

The change from 1960 (left) to 1975 (right) in the spatial distribution of suspended sediments in the Atchafalaya/Terrebonne study area. The region is 60 kilometers square, bounded on the north by the intercoastal waterway, on the west by the Atchafalaya River, and on the south by the Gulf of Mexico. The colors shown in the color bar represent the predictions of the CELSS model for each 1-kilometer zone. The change from orange to green and from purple to pink denotes an increase in suspended sediments, caused by the construction of the levee system, that accelerated the loss of wetlands.

Costanza, Sklar, and White to predict wetland loss and ecosystem succession for coastal Louisiana. The area to be modeled is covered by 2479 cells, each measuring one square kilometer. Each cell mathematically exchanges water and suspended materials with its four nearest neighbors. The suspended material consists of salts, nitrogen, and organic and inorganic sediments; it is the distribution of this material at the end of the simulation that determines whether the wetlands have gained or lost land. When the physical conditions in a cell have become appropriate, the simulation notes the introduction of a different ecosystem type.

The model takes into account a variety of factors that influence the final distribution of the suspended materials. Costanza and his team specify the weekly flow rates of the inflowing rivers and their sediment and nutrient content;

the sea level and salinity of the Gulf; and rainfall, wind, runoff, and temperature. The interaction of these factors is captured by the linking together in each cell of a set of variables that includes temperature, biomass production, salinity, water flux, and nutrients.

To verify the model, the scientists simulated a twenty-two-year period that began in 1956, by running the model on the Cray X-MP at the National Center for Supercomputing Applications. During that period, the levee systems channeled river sediments farther downstream, causing the loss of wetlands in freshwater swamps and marshes. The model captured roughly 90 percent of the changes noted in field observations. The verified model has since been used to study the outcomes of a wide variety of management proposals for the area, including the Corp's levee extension and the rises in sea level from global warming. It was found that the levee extension would create the loss of fewer than 5 square kilometers of wetlands, mostly of brackish marsh. Surprisingly, a rise in sea level of the amount projected to result from global warming actually increases the wetland area.

The Global Change Initiative

A series of studies by the National Research Council have identified several fields of research as most immediately important to reducing scientific uncertainties about global change. The Council proposes studies of the role of clouds in climate, studies of the fluxes of important chemicals (carbon, sulfur, nitrogen oxides, methane, CFCs) in the oceans, atmosphere, and biosphere, long-term measurements of the structure and function of ecological systems, the establishment of a reliable paleoclimate record, the development of comprehensive databases recording human interactions with the environment, studies of coastal erosion, and the monitoring of the incoming ultraviolet flux from the Sun.

The U.S. government has adopted these research goals as part of its plan to establish a scientific basis for policymaking aimed at addressing changes in the global Earth system. To achieve its plan, the government will fund the gathering of data from the field, the basic research to explore the processes that influence the Earth, and finally, the further development of supercomputer models of the Earth system. Two crucial components of this program are a comprehensive system for continuously monitoring the Earth and models that capture the human contribution to global change.

The first component will be primarily managed by NASA through its program Mission to Planet Earth. A major element of this program will be an Earth Observation System of more than a dozen satellites to be launched over the next fifteen years. These satellites will be generating several trillion bytes of data every day by the turn of the century. A variety of supercomputer architectures will form the core of the Earth Sciences Distributed Information System, which will process the data gathered by NASA, archive it at a series of scientific research centers, and produce analyses such as the changes through time in ocean temperature, stratospheric ozone levels, or the amount of forest and desert. The analysis of such detailed observations will require the development of incredibly sophisticated supercomputer models. Up to 10,000 scientific researchers will use the national network to access the data, analyze it, and compare it with supercomputer simulations.

No matter how detailed, the physical, chemical, geological, and biological models will not be able to give us firm predictions of global change unless we develop equally detailed descriptions of the contributing human activities. A recent, pioneering attempt to create such a model takes its inspiration from a 1987 report of the U.N. World Commission on Environment and Development entitled *Our Common Future*. The report has been widely praised for outlining strategies for managing the conflicting goals of increasing the prosperity of all human beings while protecting the environment.

In 1992, a major study by Faye Duchin, director of the Institute for Economic Analysis at New York University, and her colleagues translated the general strategies discussed in *Our Common Future* into a computer model of the world economy. The model represents the nations of the world as 16 regions. Large tables track over 50 different fuels, raw materials, and manufactured products as they flow into and out of these regions. Financial trade-offs for each region are tracked as countries decide which mix of consumption, investment, and international trade and credit to pursue.

Duchin's study compares the strategy of investment in more efficient and less polluting energy sources suggested by *Our Common Future* with a continuation of "business as usual." It finds that although both approaches have the world gross domestic product increasing by 150 percent in the next thirty years, the path suggested by *Our Common Future* allows more of that increase to go to individual consumption, by lessening the amount spent on the energy sector. The study also calculates how emissions of the oxides of carbon, sulfur, and nitrogen grow in response to a region's choice of energy sources. Unfortunately, even though the proposals of *Our Common Future* dramatically slow the use of fossil fuels, emissions still increase by 73 percent during the next thirty years. The percentage of global carbon dioxide emissions from the rich, developed nations drop to only one-third of world emissions by the year 2020 although the absolute emission amounts increase, a result that points to the need for the Earth's nations to negotiate energy and pollution trade-offs.

It is reasonable to expect that social science models will become critical to negotiators wishing to compare detailed trade-offs. Gradually the models will expand to include not only individual countries, but also societal structures such as population distribution, markets, distribution channels, political and economic systems, and the spreads of incomes and educational levels. These models would seem to naturally map onto the large number of independent processors envisioned for the next generation of massively parallel supercomputers. Eventually such sophisticated approaches will be able to integrate the driving forces of human populations with the physical models of the Earth system, creating a mathematical system that can help humankind achieve a better understanding of the possible futures that we may create for our planet.

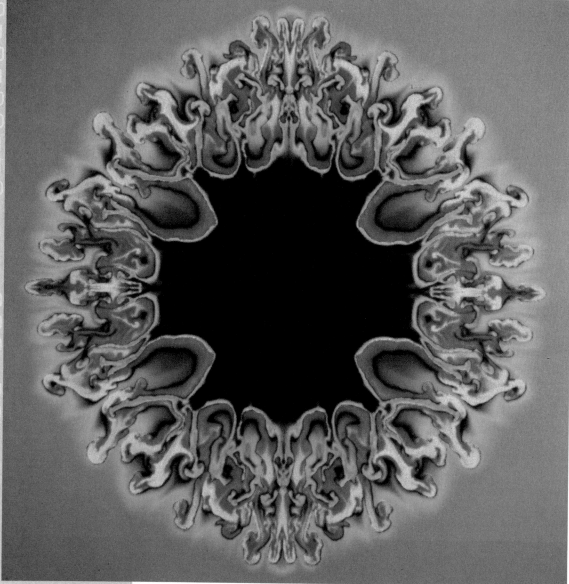

Discovering the Universe

8

Astronomy is one of the oldest fields of supercomputing applications. Stellar evolution codes have been using new supercomputers since at least the early 1950s. Many phenomena from high-energy astrophysics, such as supernovae explosions or shock wave interactions, were pioneered by researchers in the nuclear weapon design laboratories in the 1960s and 1970s. In the late 1970s, the Einstein Observatory x-ray telescope and the Very Large Array of radio telescopes created images using computers as intermediaries between the telescope and the observer. With the access provided in the late 1980s by the national network and the National Science Foundation supercomputer centers, the academic astronomical community has become a major participant in computational science.

Deep inside a doomed star, the helium-burning shell mixes into the overlying hydrogen, four hours after a supernova explosion begins to tear the star apart. The "mushroom" features are characteristic of Kelvin-Helmholtz instabilities building on top of the finger-like Rayleigh-Taylor instabilities. Color varies from blue for low densities, through red, to yellow for the highest densities. The dark area in the center represents the core, which is composed of heavier elements.

In the 1990s, high-performance computing and communications will continue to transform the practice of both observational astronomy and theoretical astrophysics, as scientists bring together on their desktops realistically complex simulations performed on teraflop supercomputers and multi-wavelength high-resolution observations obtained over the network from national digital archives.

We will start our journey where we left off in the last chapter, with the planet Earth, and work our way through the solar system, out to the stars, and beyond to galaxies and clusters of galaxies. Yet as grand as all this appears, we will find that all the glorious luminous matter in the universe is dwarfed by unseen dark matter that determines the large-scale structure of the universe.

The Planets

One of the classic questions in astronomy concerns the origin of our Moon. Traditional explanations, like the hypothesis that the Moon spun off from a rapidly rotating primordial Earth, or that it was created elsewhere in the solar system and then captured by the Earth's gravity, conflict with analyses of the moon rocks brought back by the Apollo astronauts. These rocks are quite like deep-seated material from the Earth's mantle, except that they are depleted of volatile substances as if they had been baked at a high temperature. Whereas Earth rocks contain water, moon rocks are bone dry. If the Moon had broken away from a spinning, molten Earth, the moon rocks should be nearly identical to Earth rocks. Conversely, if the Moon had formed elsewhere in the solar system, lunar abundances of certain isotopes should be noticeably different from terrestrial abundances.

In the mid-1970s, two research teams independently proposed an innovative theory to account for the Moon's creation. Perhaps the pri-

mordial Earth was hit by a planet-sized object, possibly as large as Mars. The violence of such a collision would have ejected a huge amount of material from the Earth's mantle, and some of this material would have entered Earth orbit. Eventually this orbiting material would have cooled and coalesced to form the Moon. Supercomputer simulations afford an excellent means of studying how reasonable this collision-ejection hypothesis really is.

One of the first simulations of a Mars-sized object obliquely striking the Earth was performed by Marlin Kipp at Sandia National Laboratories and Jay Melosh at the University of Arizona. They used a code that was especially designed to handle shock waves and material deformation accurately. As materials are subjected to forces induced by the impact, finite difference algorithms calculate the transport of mass according to the conservation laws for mass, energy, and momentum.

The Earth and the Mars-sized projectile are each assigned an iron core and a silicate mantle. Details of the behavior of these materials are obtained from so-called equations of state. These equations also properly account for phase changes—the conversion of solid to liquid and liquid to vapor—that certainly would occur in some locations during a collision of two planets. In the simulation shown on the facing page, the cores and mantles of the Earth and the impactor are color-coded to make it easy to follow material from different parts of the two worlds as the collision proceeds. The impactor strikes the Earth at a speed of 12 kilometers per second, typical of relative velocities we find in the solar system. Following impact, an enormous plume of vaporized rock erupts from the Earth's mantle.

Kipp and Melosh used a code that accurately simulated the shock phenomena produced by the impact, but it could not follow the plume of debris into orbit as precisely. One problem is that gravity in the simulation was directed only toward the center of the Earth, so the ejected rock did not itself have any gravity—a deficit

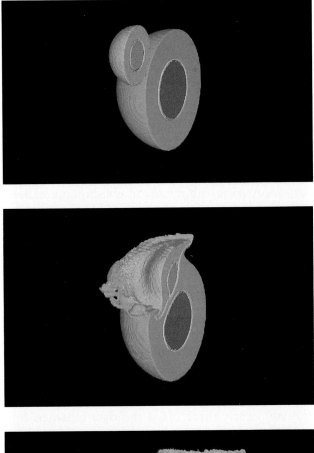

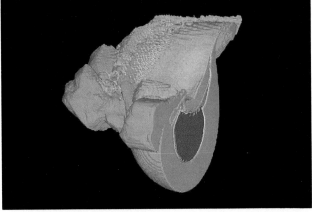

The Earth is struck by a projectile of nearly the same size and mass as the planet Mars in these three images from a numerical experiment exploring the origins of the Moon. The first view shows the moment of impact; the remaining views show the consequences 800 and 1600 seconds after impact.

that must eventually cause the simulation to become unrealistic. Nevertheless, it is clear that the high temperatures of the ejecta would have easily driven off volatiles, while leaving isotopic abundances unaltered.

To simulate the behavior of the ejected plume more accurately, Willy Benz at the Harvard-Smithsonian Center for Astrophysics and his colleagues turned to the method of smoothed particle hydrodynamics. This method applies particle methods to a continuous fluid by replacing the fluid with a set of spatially extended particles. The simulation evolves with time by computing the trajectories of all the particles from the various forces that act on them. These forces depend on the distances between the particles and on their relative velocities. *N*-body methods are used to compute the gravitational component of the forces, and for this reason, tree codes often help make the problem tractable. Because a conventional mesh is not needed, smoothed particle hydrodynamics is well suited for the simulation of highly distorted flows, such as the collision ejecta, which would otherwise cause a conventional mesh to become tangled or severely misshapen.

For their simulation, Benz and his colleagues modeled the Earth and the impactor with 3000 extended particles. Using a Cray X-MP at Los Alamos National Laboratory, they examined various collision scenarios to see which are most likely to produce the Moon. The proto-Earth and the impactor were given just slightly more angular momentum than the Earth-Moon system has today, thereby allowing for angular momentum removed from the system by escaping particles. The scientists varied the mass ratio of the bodies for each scenario and inspected resulting simulations for the formation of a large Earth-orbiting satellite. They found that optimum conditions for the Moon's formation were obtained when the mass ratio of the impactor to the proto-Earth was 0.136. For comparison, the present ratio of the mass of the Moon to that of the Earth is 0.012, while the ratio of Mars's mass to Earth's is 0.107.

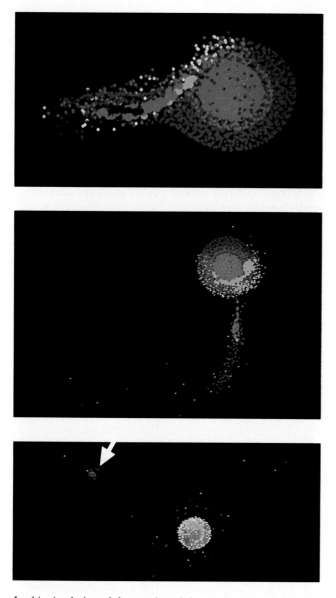

In this simulation of the creation of the moon, the grazing impact of a Mars-sized object (coming from the right) on the Earth ejects both mantle and core material. Most of the iron from the core falls back onto the Earth, and the surviving ejected rock material coalesces to form the moon (indicated by the arrow), pictured during its first orbit. To follow ejected material as it moves away from the Earth, successive views show increasingly larger volumes of space. Blue and green indicate iron from the cores of the Earth and the impactor; red, orange, and yellow indicate rock mantle material.

Supercomputers of various architectures are being used in quite a different manner to pierce the cloud cover of our sister planet Venus. The *Magellan* spacecraft carried aloft by the Space Shuttle *Atlantis* on May 4, 1989, now orbits Venus seven times each Earth day, producing images of the surface in strips 20 kilometers wide. It resolves details as small as 120 meters, about the size of a football field. Because the thick cloud cover over Venus makes direct optical imaging useless for viewing all but the upper atmosphere, *Magellan* was designed to see with radio frequencies by means of a technique called synthetic aperture radar. Since the Venusian atmosphere is transparent at these wavelengths, the spacecraft can sweep the planet's rocky face, measuring the radar backscatter to determine details of the surface. An altimeter independently maps the topographic relief of the planet.

Back on Earth, the continuous data stream is collected at the Jet Propulsion Laboratory in Pasadena, California. A complex network including both workstations and supercomputers acts as a metacomputer to produce a wide range of images for scientific study. The volume of data returned by *Magellan* is almost one terabyte, nearly three times the data returned by all previous U.S. planetary missions combined. The task of converting the raw data into images is so repetitive that the Jet Propulsion Laboratory has created a special computer to perform that function only. The Digital Correlator Subsystem is one of only two or three similar systems in the world. Because its image- and signal-processing algorithms are hard-wired into a pipeline that processes the dataflow, the Digital Correlator System can sustain speeds of 6.4 gigaflops. The machine is combining the altitude data with the surface detail mapped by synthetic aperture radar to build up a database of planetary features that covers 90 percent of the surface.

The Multimission Image Processing Laboratory at the Jet Propulsion Laboratory is able to convert selected portions of this database into animated images that can re-create a virtual trip over the Venusian surface from any angle. The 512-node Intel Delta and a Cray Research supercomputer speed the production of breathtaking representations of volcanos, mountains, and impact craters, colored to match surface colors detected by the Soviet Venura landers. A videotape takes the viewer on a roller coaster ride over the surface that will probably never be duplicated by human beings actually traveling on the hostile planet, whose temperature exceeds 900°F and whose air pressure is almost 100 times the Earth's. Besides providing a way for the public to share in the excitement of discovering a new world, these virtual trips give planetary scientists a global context in which to fit their detailed studies of particular surface features.

Magnetospheres

A plasma gas of electrons, protons, and charged atoms streams from the Sun in a solar wind, which carries with it the interplanetary magnetic field (IMF). When the IMF reaches a planet having its own magnetic fields, it confines the dipole magnetic fields into a volume with a long teardrop shape, called a magnetosphere. Although we normally regard Jupiter as the largest object in the solar system besides the Sun, it is dwarfed by the scale of the magnetospheres of the planets. The defining magnetic field lines of these enormous structures are constantly changing in strength and location, forming a dynamical structure that has gradually been deciphered

This still frame from an animation of a virtual ride over the surface of Venus shows Maat Mons, the largest volcano on Venus. The view is taken from an imaginary observer stationed 3 kilometers above the surface, 634 kilometers north of the volcano. The volcano is over 5 kilometers high, comparable to Mauna Loa on Earth. The surface appears orange because the thick clouds absorb the blue components of sunlight. The vertical scale has been exaggerated 10 times.

over the last three decades by a combination of planetary probes, theoretical models, and super-computer simulations.

The details of the shifting magnetosphere are difficult to model because of the wide variety of time and space scales involved. Particles of each charged species can differ widely in velocity and energy from one time to the next at each point in space. The electromagnetic field not only has a large-scale, constantly shifting structure, but it also can support a wide range of wave motions induced by the constantly fluctuating solar wind. Finally, the particles and the fields have a bewildering array of possible interactions. The goal of creating a "general circulation model" for the magnetospheric "weather" will occupy researchers well into the next century.

Nonetheless, much progress has been made by isolating certain phenomena for study or modeling only the dominant processes. We shall illustrate the rewards of these strategies by briefly reviewing three supercomputer simulations, one at the global scale of the entire magnetosphere, one at a much smaller, but still macroscopic scale, and one at the microscopic scale of the plasma constituents themselves.

An approximation to the physics called magnetohydrodynamics (MHD) is often used to simulate the macroscopic structures in magnetospheres. It assumes that the charged particles must cling to magnetic field lines, orbiting the lines in tight circles or helices. Typically, the model uses finite differencing techniques to solve the MHD equations in a manner similar to that used to study the jet flows in Chapter 3. Because the MHD approximation does not represent small-scale behavior well, scientists desiring to examine the interactions of particles and field lines turn to plasma particle codes similar to those used to model the fusion reactors discussed in Chapter 6.

As the solar wind impacts the Earth's magnetosphere, the magnetic field lines of the IMF sweep over the closed lines of the magnetosphere. Some of the open field lines of the IMF

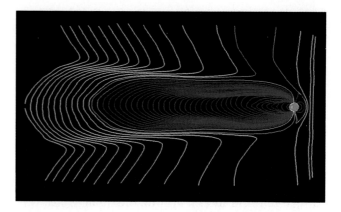

In this image from a simulation by Fedder, Mobarry, and Lyon, the open magnetic field lines (yellow) of the solar wind kink as they sweep over the Earth's (green sphere) closed geomagnetic field lines (magenta), with some reconnecting (blue) to become part of Earth's magnetosphere.

break and "reconnect" with the system of closed lines. Through these connections, the solar wind feeds energy and momentum into the Earth's magnetosphere. The strength of the reconnection between the field lines of the IMF and those of the Earth's magnetosphere depends very much on the angle between the connecting field lines.

Joel A. Fedder and Clark M. Mobarry of the Naval Research Laboratory and John G. Lyon of the University of Iowa have performed a series of global magnetosphere simulations with an MHD code on the Cray X-MP at the Naval Research Laboratory in order to map this angle-dependent relationship. For each simulation they vary the angle at which the incoming IMF hits the Earth's magnetosphere and then compute the amount of reconnection. From the results of this work, solar and magnetospheric researchers can make better predictions about solar wind-induced electrical disturbances in the Earth's ionosphere, the ionized high-altitude portion of the atmosphere that can conduct currents. These disturbances can disrupt radio, television, telephone, and electrical power service, and in extreme cases permanently damage satellites in low Earth orbit.

In stark contrast to the Earth, Venus has no magnetic field, so the solar wind slams directly into the ionosphere. Space probes have detected "ropes" of plasma attached to magnetic field lines in the vicinity of Venus. In an effort to understand how they might form, Robert Wolff, then at the Jet Propulsion Laboratory, collaborated with Michael Norman of the National Center for Supercomputing Applications to simulate the interaction of the solar wind with the Venusian ionosphere. On the side of Venus 90 degrees away from the impact point, the solar wind is sweeping past the ionosphere's surface at a tangent. If one zooms in on a small portion of that surface, one can approximate the interacting wind and atmosphere with a two-dimensional simulation analogous to the Kelvin-Helmholtz instability that we saw Woodward compute in Chapter 1, except that the fluids are now magnetized.

The upper part of the grid is filled with a magnetized, hot (100,000 K), tenuous solar wind plasma moving to the right at 100 kilometers per second. The bottom part of the grid is the unmagnetized, dense, cold (5000 K) plasma at rest representing the Venusian ionosphere. After running their model on the NCSA's Cray X-MP for ten hours, Wolff and Norman found that small-amplitude perturbations on the surface between the two plasmas grow quickly, until the solar wind is taking a huge "scoop" out of the ionosphere. The magnetic field carried by the solar wind becomes rolled up with dense cold plasma from the ionosphere, forming structures that resemble the observed ropes.

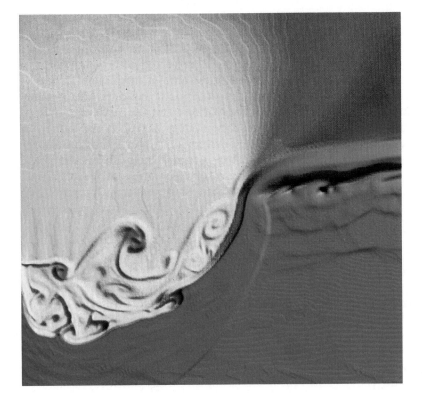

Strong waves roll up as the magnetized solar wind, sweeping across the top of the image from left to right, digs down into the Venusian ionosphere, pictured at the bottom. A color map representing density is painted over the shaded relief, whose height represents the value of the pressure. The smoothly curved blue ridge represents a shock wave.

Finally, there is a very large class of microscopic computations that aim at understanding the basic interactions of the particles and magnetic fields. A good example of this type of calculation is found in the work by Maha Ashour-Abdulla and David Shriver at the University of California at Los Angeles. Using particle codes on the San Diego Supercomputer Center's Cray Y-MP, they found a new mechanism to explain the broadband electrostatic noise that spacecraft pick up when passing through the tail of a magnetosphere. Although it was already known that the interaction of ion beams with the plasma in the solar wind can create the noise, they found that currents, generated by electrons beaming through the tail and aligning themselves with the magnetic field, can produce the noise in the absence of any ion beams.

The Life Cycle of Stars

In the spiral arms of the Milky Way galaxy, stars are continually being born from the condensation of interstellar gases. "Nurseries" of newborn stars are just one interesting feature of the spiral arms, which also contain a jumble of huge dust clouds formed of many different organic molecules, shock waves originating from supernovae, and "protostellar" jets from stars that are just being born. The Hubble Space Telescope is already providing us with spectacular high-resolution pictures of some of these events.

In the 1990s, a large variety of new Earth- and space-based telescopes will be observing these regions as well. Only wavelengths longer than light can penetrate the extensive dust in star-forming regions. Thus the new instruments that will most greatly aid these studies all observe in the infrared: an eight-meter infrared-optimized telescope on Mauna Kea in Hawaii, a forty-dish Millimeter Array, and the Space Infrared Telescope Facility.

These telescopes will come into existence at about the same time as teraflop computers and the gigabit per second National Research and Education Network. The question arises, could the synthesis of the new telescope technology and the new information technologies enable new modes of observation by the end of the decade? To obtain some insight, the National Center for Supercomputing Applications is working with the Berkeley-Illinois-Maryland Array of radio telescopes in northern California to explore novel ways of carrying out observational astronomy. BIMA, which observes in the millimeter wavelength range, is connected to the NCSA metacomputer by a transcontinental high-speed network.

BIMA will be able to detect incoming millimeter signals at 1024 frequencies simultaneously. An observation produces a "data cube" of two spatial dimensions on the sky and one dimension of frequency. Each data cube is computed from the correlated inputs of all the dishes, traditionally days to months after the data are taken.

If a supercomputer is attached to the array by a high-speed link, an astronomer connected with a nationwide network will be able to guide the observations from a desktop computer as he or she sees the data converted into visual form by the supercomputer almost immediately. If problems are detected, or if the data are of poor quality, the observations can be made again while the telescope system is still in the same configuration. Conversely, if something unexpected or startling is found, new observations can begin immediately. After several years of experimentation with this new form of remote observation, dedicated computers will be attached to the telescope to perform imaging and respond to the astronomer's instructions.

Astrophysicists will require teraflop computers to model the formation of a star at the level of detail necessary to interpret the wealth of observational data on star-forming regions that will flood the astronomy community over the next decade. The models must include equations to handle gas dynamics, magnetic fields, radiation coupling, heating and cooling, and still more processes. Furthermore, since most stars are formed in binary systems, the processes must occur in a highly nonspherical geometry that is correspondingly difficult to model. Certainly, the understanding of star formation should be one of the major "grand challenges" of the 1990s.

After formation, a star settles down as a roughly spherical ball of luminous gas powered by nuclear reactions at its center. The mass of the overlying layers crushes down on the interior of the star, giving the core an extremely high temperature and pressure. For example, the temperature at the Sun's center is estimated to be about 15 million K, while gases at the solar surface measure only 5800 K. Once newborn stars begin burning hydrogen in their cores, they can live anywhere from tens of millions to tens of billions of years. The lifetime of the star is governed by its total mass; the more massive the star, the hotter the central core becomes and the faster the nuclear fusion reactions exhaust the hydrogen fuel.

For over forty years, supercomputers have assisted studies of stellar evolution. Until recently, they were not fast enough to compute any geometry more complicated than a sphere. The supercomputer keeps track of the state of the gas in each of the star's radial shells, calculating the gas's density, temperature, and pressure. It also follows the network of nuclear reactions within each shell that describes which element is undergoing nuclear fusion and which are being generated as "ashes," and it tracks the flux of heat and radiation passing outward.

To understand the structure and evolution of a star, it is necessary to know exactly how energy is transported outward from the star's center to its surface. In normal stars, energy flows outward by two different mechanisms. If the gases inside the star are sufficiently transparent, photons can travel substantial distances before colliding with atoms, thereby directly conveying energy from one location to another. This process is called radiative diffusion because photons gradually migrate outward from the warmer interior toward the cooler stellar surface.

If, however, the gases in a star are sufficiently opaque, photons cannot travel very far before they are absorbed by atoms. The absorption of photons creates hot regions that become buoyant and rise, carrying the heat outward by convection, just as occurs in the Earth's mantle.

However, unlike the convection inside the Earth, the density varies by a factor of 10,000 between the top and bottom of the convection zone, which occupies the outer one-third of the Sun's radius. In some stars, convection can penetrate to even deeper layers. It can transport fresh fuel downward into the thermonuclear inferno at a star's core, or it can dredge up newly created elements and spew them into the star's atmosphere. This mixing has a dramatic effect on the star's appearance and evolution.

Paul Woodward and David Porter at the University of Minnesota have used the Pittsburgh Supercomputing Center's Cray Y-MP to model convection in the Sun in two spatial dimensions, over a large number of grid zones. Their computational space has two layers, a buoyantly unstable gas layer on top and a stable gas layer underneath. A higher temperature is maintained along the bottom and a lower temperature along the top. In response to the downward pull of the star's gravity, the density and pressure of the gas decrease rapidly from a maximum along the bottom of the grid to a minimum at the top.

Each plume of cool gas descending into the star is usually accompanied by two vortices spinning in opposite directions at speeds close to the speed of sound. The rotation is so rapid that the density at the center of a vortex is reduced to a tenth of the ambient fluid density. Furthermore, vortices spinning in the same direction tend to merge while those spinning in opposite directions generally avoid each other. Sound waves created by these vortices steepen into shock waves in the low-density fluid near the top of the grid and then rattle around inside the simulated star.

Woodward and Porter discovered that occasionally a descending vortex pair overshoots

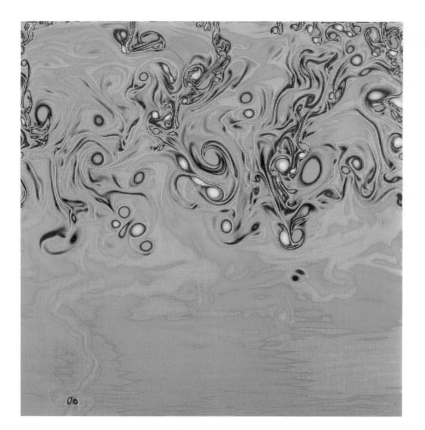

the unstable zone and penetrates the stable zone below. However, because the two vortices are now in a stable layer, they do not bob back up into the unstable layer. This discovery provides a specific mechanism for a process that had been theoretically suspected before the simulation. Namely, that by this mechanism of "penetrative convection" fresh hydrogen nuclear fuel can be furnished to the denser core where hydrogen is fusing. The provision of fresh fuel would lengthen the time the star sits on what astronomers call the "main sequence," a classification of stars comprising those that, by their surface temperatures and luminosity, are concluded to be burning hydrogen in their cores.

Using the highly accurate algorithms developed by Woodward and Phillip Colella of Berkeley, Andrea Malagoli, Fausto Cattaneo, and Robert Rossner of the University of Chicago have performed three-dimensional simulations of the Sun's convection on the Cray Y-MP at the Goddard Space Flight Center. They find that the upper part of the convection zone has a granular pattern similar to that seen on the Sun's surface, whereas near the bottom the structure of the flow changes completely, because of the large density differences. The change in structure strongly affects the mechanism by which the total energy is transported to the surface.

Although these are useful beginnings, the current generation of supercomputers is not powerful enough to model stellar convection in its full complexity. The three-dimensional simulations need to be carried out with the fine resolution of Woodward and Porter's two-dimen-

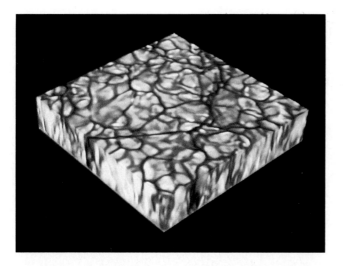

This image from Malagoli and Cattaneo's three-dimensional simulation shows typical convection patterns that occur under circumstances like those in the upper layers of a star. The granularity, representing individual convective cells, is similar to that seen on the Sun's surface. The colors represent temperature, while intensity indicates velocity, with downward-moving gas being dark and upward-moving gas white.

sional simulations. Moreover, in a real star some of the violent turbulent motions are probably inhibited severely by magnetic fields, and the inclusion of these fields would dramatically lengthen the time needed for the simulation. Thus, these and other researchers are eagerly awaiting the teraflop supercomputers that will allow their models to achieve higher resolution and to consider the additional physics needed to compute conditions in real stars.

Although most stars evolve peacefully for billions of years, more massive stars live more brilliantly but more briefly, ending their shorter lives in a violent manner. On February 23, 1987, a blue supergiant star, about 100,000 times as luminous as our Sun, ended its life in a supernova explosion. This supernova, termed SN1987A, took place in the companion galaxy to our own Galaxy, called the Large Magellanic Cloud. It was the brightest supernova seen from Earth since 1604, when a supernova was ob-

served with the crude instruments of the time by the astronomer Johannes Kepler. In contrast, supernova SN1987A was studied in detail by almost all modern observing devices on Earth and in orbit.

Before this event, most supercomputer simulations of stellar evolution showed massive stars ending their lives in supernovae while they were in the red supergiant stage. Astronomers such as Stanley Woosley and Thomas Weaver of Lick Observatory and the Lawrence Livermore National Laboratory and David Arnett at the University of Arizona improved older stellar evolution calculations by treating more carefully both elemental abundances and the coupling of the stellar core to its atmospheres. Once they took into account the lower amounts of heavier elements found in the Large Magellanic Cloud compared to our Galaxy, their simulations showed that a star with 18 to 20 times the mass of the Sun would naturally pass through the red supergiant phase to the blue, eventually becoming a supernova in the blue phase.

Their computations showed that the star would have burned hydrogen in its core for about 10 million years before exhausting that fuel, compared to the approximately 10 billion years it takes for a star of our Sun's mass. After exhausting its hydrogen, which would have been turned into helium by the fusion process, the core contracts to a density 200 times greater until the temperature becomes high enough to support helium fusing. At the same time the atmosphere expands, turning the star into a red supergiant.

This stage of helium burning lasts for another 600,000 years. As the core helium becomes exhausted, the star contracts, becoming a blue supergiant. Then carbon burns in the core for 12,000 years, followed by neon for 12 years, oxygen for 4 years, and finally silicon, which burns to iron, for 1 week. Iron can no longer produce energy by fusion, so the iron core, now 1.5 solar masses, is doomed to collapse in a fraction of a second to the density of nuclear matter. Outside this iron core, the remaining 5

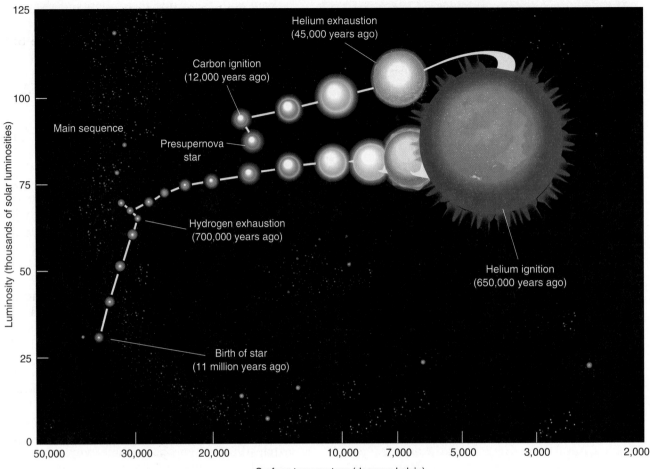

A star of 18 solar masses in the Large Magellanic Cloud would evolve through the stages illustrated by this chart, which gives the relative size, color, surface temperature, and luminosity of the star at each stage. At its largest, the red giant phase, the star's outer radius exceeds the distance from the Earth to the Sun.

solar masses of helium and heavier elements continue to burn slowly, and each of the fusing elements remains isolated in one of a series of spherical shells stacked like the layers of an onion. Another 10 solar masses form a hydrogen atmosphere surrounding the shells of actively fusing elements.

This distribution of density and elemental abundance becomes the initial data for another class of supercomputer codes, designed to model the collapse of the iron core. These codes, such as the one developed by James Wilson at Lawrence Livermore National Laboratory and his collaborators, must handle shock physics, nuclear physics, neutrino transport, and the relativistic motion of matter. Because these physical effects are so delicately balanced, there is still no consensus on the quantitative details of how the core collapse creates a supernova, even though supercomputer models have been exploring the creation of supernovae for more than twenty-five years.

Nonetheless, it is clear from these supercomputer simulations, as well as from theoretical studies, that when the imploding core hits nuclear density, the Pauli exclusion principle comes into play and the collapse suddenly halts. The onrushing matter is abruptly decelerated by the standing core, and a very strong shock wave races back out through the overlaying star. An enormous burst of neutrinos is generated, whose luminosity for a millisecond exceeds the entire optical luminosity of the observable universe!

Some of the momentum of the outgoing neutrinos helps the shock to lift off the layers of the star. As the stellar layers move outward, the central core of the star forms a very hot young neutron star. A neutron star is composed mostly of closely packed neutrons compressed by gravity into a sphere about 30 kilometers in diameter. The initial burst of neutrinos from SN1987A was detected on Earth by huge underground "neutrino telescopes" in Japan and the United States, and its detection confirmed the basic model that had been developed by supercomputer modeling over the previous several decades.

As time passed, the decaying light curve of the supernova agreed with the light curve predicted by the onion skin model that had emerged from the stellar evolution computations. However, some intriguing discrepancies brought into question the assumption of spherical symmetry that had allowed codes to model both the core collapse and the stellar evolution in only one dimension. Observations of the supernova in gamma ray, x-ray, optical, and infrared wavelengths began to build up evidence for mixing between layers of the core. Of course, mixing is impossible in the neatly stacked spherical model. Realizing that nonspherical instabilities must be causing the mixing, groups in Japan, Germany, and the United States turned to supercomputers to run new codes that could handle the two or three spatial dimensions needed for this much more difficult case.

Great care had to be taken to ensure both that the spatial grid was fine enough to follow the mixing "fingers" and that the starting model of the spherical stellar core was accurate. A stunning result from a simulation that met both requirements is shown on page 196, from work by Bruce Fryxell and David Arnett of the University of Arizona and Ewald Müller of the Max Planck Institute for Astrophysics in Munich, Germany. Using a Cray-2 at the Rechenzentrum der Universität Stuttgart and a Cray-2 and a Cray Y-MP at the National Center for Supercomputing Applications, the scientists have followed small nonspherical perturbations as the layers begin to expand in the explosion. To resolve the fine detail, they relied on both a fine grid of one million zones and the highly accurate algorithms developed by Woodward and Colella.

When a glass of water is turned upside down, the situation is unstable because heavy water is sitting on light air in a downward-pointing gravitational field. The instability that eventually breaks the water blob into droplets is called a Rayleigh-Taylor instability. Similarly in an exploding star, the heavy layers become unstable from the outward push of the shock and begin to develop long fingers of denser material that mix with the lighter overlying material.

The helium layer mixes outward into the hydrogen layer, the oxygen layer mixes outward into the helium layer, and so forth, all the way down through the "onion." As a result of these simulations, astronomers can now begin to compare in detail observations of the velocities of the different elements emerging from SN1987A with the dynamics of the mixing displayed in the simulation. Any discrepancies may hint at even more unusual phenomena, such as a highly nonspherical stage during the core collapse.

As the shock wave from the explosion moves outward at 10,000 kilometers per second, it ultimately comes in contact with the interstellar gas or with the outer envelopes of the star that were shed in early phases of stellar evolution. At that point, the shock "snowplows" into the gas, strongly heating it and causing it to emit electromagnetic radiation, often in wavelengths from x-ray to optical to radio. Five years after the supernova explosion, the shock wave from SN1987A has not yet traveled far enough to reach this stage. To obtain a glimpse of this future state, we can look at the supernova remnant, called Cassiopeia A, created by the outgoing shock of a much older supernova that exploded 300 years earlier. Oddly enough, when the actual supernova explosion occurred, less than one hundred years after the one observed by Kepler, no one seems to have observed it.

Even today, the Cassiopeia A remnant is very hard to detect with optical telescopes, de-

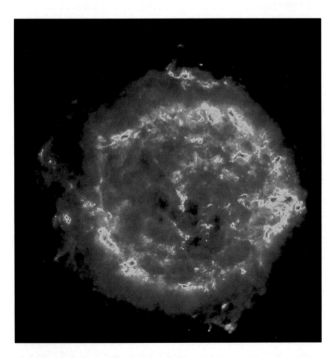

A high-resolution radio image of the supernova remnant Cassiopeia A, the result of a supernova explosion observed three centuries ago. The colors are proportional to the intensity of the radio emission, with red and yellow representing the highest intensity.

spite the large amounts of gas piled up in front of it by the shock. In less than an hour, however, a Cray X-MP supercomputer was able to produce an extremely high-resolution radio map of the supernova remnant from observational data obtained with the Very Large Array of radio telescopes. The image's 4000×4000 pixels capture far more detail than the roughly 640×480 pixels on most personal computer screens. The features seen in red and yellow on this high-resolution image once again demonstrate that although stars are primarily spherical through most of their lives and death, deviations from sphericity may be important.

Stars are born in groups, and the stars within a group are born with a wide range of masses. Since the most massive stars burn out first, a sequence of supernovae may explode in a region of a galaxy. The outward force of the explosions pushes back the gases of the galactic atmosphere, forming a bubble containing rarified matter inside and compressed gases along the surface. The remaining brilliant massive stars emit fast-moving stellar winds that enlarge the bubble in the galactic atmosphere. Because the spiral galaxy is very thin, the bubble will eventually poke out the top (or bottom) of the spiral arm. The pressure on its top will then be considerably lower than the pressure on its sides. At that point the bubble bursts out into the upper reaches of the thinner galactic atmosphere.

Mordecai-Mark Mac Low and Richard McCray of the University of Colorado collaborated with Michael Norman of the National Center for Supercomputing Applications to model the formation of superbubbles in atmospheres of differing density profiles. Using a gas dynamics code on NCSA's Cray X-MP, they found that most of the swept-up mass, compressed by the expanding bubble, stays in the plane of the galaxy, but that the breakout vents hot gas into the halo of the galaxy. They conclude that a galactic halo may exist consisting of a froth of merged superbubbles created in star-forming regions throughout the galaxy.

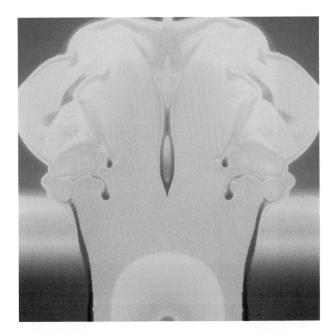

A superbubble in a galactic atmosphere expands nearly spherically until its top begins to feel lower atmospheric density, at which point it blows hot gas out into the galactic halo. The falloff in density is indicated by the horizontal stripes of color, which trace the contours of the logarithm of the gas density. The yellow fingers are Rayleigh-Taylor instabilities similar to those seen on page 196.

Colliding Compact Objects

What happens to the collapsed cores of supernovae? Much depends on whether the exploding star is a single star or a member of a binary star system. Double stars are so common that they constitute about half of the stars you see in the night sky. They also include some of the most intriguing stellar systems known to astronomers because of a multitude of exotic phenomena that can occur when two stars evolve in each other's proximity. For instance, it is possible that both stars in a binary system are sufficiently

massive to become neutron stars at the end of their lives. The one such known system, called PSR 1913+16, consists of two neutron stars revolving about each other in only $7\frac{3}{4}$ hours, at a separation of about 2 million kilometers, only a little larger than the diameter of the Sun!

Astronomers are interested in binary neutron stars because they should be strong sources of gravitational waves, the ripples in the fabric of space caused by a changing nonspherical gravitational field. Einstein's general theory of relativity predicts that as two neutron stars revolve about each other, they lose orbital energy through the continuous emission of gravitational radiation. Because of this energy loss, the two stars spiral toward each other and ultimately collide and coalesce. Such a collision should release a powerful burst of gravitational waves.

As a first step toward studying such cosmic collisions, Charles R. Evans of the University of North Carolina used NCSA's Cray X-MP to simulate a head-on collision between two neutron stars. Besides limiting the collision to a head-on case, Evans did not attempt to adjust Newton's law of gravity for the effects of general relativity, and he also ignored any heating or cooling that would result from the release of neutrinos. His grid consisted of 13,500 grid points, and the computation required nearly 10 hours of supercomputer time.

As the neutron stars fall toward each other, they continue to accelerate until their collision creates a powerful recoil shock wave that moves outward in all directions but is strongest passing back through the stars, along the axis of symmetry. The merged stars then undergo large-amplitude oscillations as the core alternately reexpands and contracts. A novel manner of visualizing the collision was created with the help of NCSA graphics specialist Ray Idazsak. The program computed the paths of a set of tracer particles that follow the trajectories of individual mass parcels.

The vigorous core oscillations produce a train of shock waves that continue to eject mat-

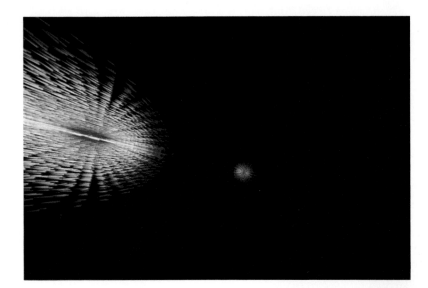

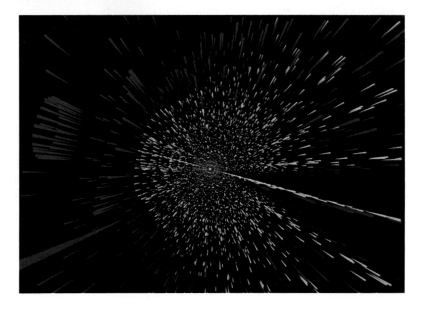

Top: *Two neutron stars speed toward impact. Red particles represent material from the inner core, yellow particles the outer core, and blue and purple particles the outer layers.* Bottom: *The collision generates an outgoing shock wave that blows off the outer material. Although there is some mixing of the core and mantle (red and yellow), the overall structure is maintained.*

ter from the system. Between 5 and 10 percent of the mass of the two stars is carried off by the shocks; the remaining 90 to 95 percent of the matter is gravitationally bound in the merged core. Applying the appropriate formulae, Evans found that about one-tenth of one percent of the mass-energy of the system is carried away by gravitational waves.

Within one year of its launch, the Earth-orbiting Compton Gamma Ray Observatory discovered bursts of gamma rays coming from more than a hundred sources scattered randomly across the sky. Colliding neutron stars could be one possible source for some of the bursts. Much more complex supercomputer simulations, which include realistic details of the spiraling and merging of two stellar corpses, will help to determine if any of these mysterious gamma ray bursts come from such collisions or from even more exotic interactions of matter with single neutron stars.

The strong bursts of gravitational waves produced by violent events cannot be observed by any instrument on Earth today. The National Science Foundation is funding the construction of a four-kilometer Laser Interferometer Gravitational Wave Observatory (LIGO) that should be able to detect these incredibly small signals by early in the next century. Even the most powerful source we can imagine seeing would alter the length of the laser beam in LIGO by only about one-hundredth of the diameter of a proton! Needless to say, the race is on to compute the expected waveforms from various astrophysical sources, with the help of supercomputers. The source most likely to be observed first is an event even more violent than the collision of two neutron stars—the collision of two black holes.

If too much mass collapses in the center of a supernova and the surface gravity of the collapsing core becomes so great that even light itself cannot escape, then rather than a neutron star a black hole will be born. According to general relativity theory, the dense accumulations of matter in a black hole severely stretch the fabric of spacetime. If the hole becomes nonspherical, it will radiate small ripples in this fabric that propagate at the speed of light. These ripples are gravitational waves.

For the last two decades, one of the authors, Larry Smarr, has been working with his colleagues on developing supercomputer codes that can solve the Einstein equations of general relativity for the distortions of space and time at a black hole. For the last five years, he has been collaborating with a group at the National Center for Supercomputing Applications that has specialized in writing codes for the study of black hole dynamics. The group has included Ed Seidel, David Bernstein, Andrew Abrahams, now at Cornell University, and David Hobill, now at the University of Calgary, Canada.

When two nonrotating black holes collide head on, they first form a nonspherical black hole. The difference between the mass of the final hole and the sum of the masses of the original holes gives the amount of energy that radiates from the collision as gravitational waves. As the black hole radiates, it gradually sheds its nonsphericity, until it finally settles in a static spherical state. Of course, some of the waves are radiated down the hole and do not escape to distant observers.

Recently the NCSA group used a Cray Y-MP to compute this conversion to sphericity. Collaborating with Ray Idazsak and computer artist Donna Cox of NCSA and the University of Illinois, the group was able to create a visualization showing both the spatial curvature of the

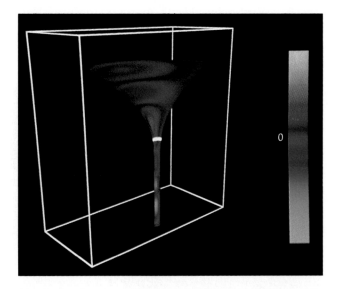

Gravitational waves emerge from a nonspherical black hole. The funnel shape shows the curvature induced in the surrounding space by the black hole, whose surface is defined with a white line. Some waves are able to just barely pull away from just outside the hole, while others go down its throat. The colors represent peaks and troughs in the gravitational waves. Colors above 0 on the color bar represent peaks of increasing height, while those below 0 represent troughs of increasing depth.

black hole and the outgoing and ingoing waves. There is now forming a national coalition of relativists and computer scientists who plan to use the coming teraflop supercomputers to extend their ability to solve the Einstein equations. They hope to solve the general case of a collision between two spiraling black holes and to compute the waveforms expected to be observed with the gravitational wave observatory.

Active Galactic Nuclei

Some of the most studied astronomical objects of the last three decades have been large-scale radio-emitting lobes and jets issuing from the centers of active galaxies, such as the jets from Cygnus A that we saw in Chapter 3. In the last decade it has been discovered that jets ending in double radio lobes, looking very similar to Cygnus A, are associated with many of the brightest objects in the universe, the mysterious quasars.

These distant powerhouses appeared only as points of light to optical telescopes when they were discovered in the early 1960s. However, the spectrum of these starlike objects was so shifted toward the red that scientists had to conclude they were at cosmological distances, billions of light years away.

Where radio jets and quasars coexist, the quasar is in the position one would assume is the galactic core. Strong observational evidence suggests that the jets emerging from many quasar cores are moving at speeds very close to the speed of light. The current theory suggests that these objects, which can be one thousand times brighter than an entire ordinary galaxy, are indeed the overly luminous cores of galaxies and that their energy is derived from gas accreting onto a massive black hole of perhaps several million to several billion solar masses. This conclusion is strongly supported by examining much closer active galaxies of much lower luminosity that have features in common with quasars. The energy to power both objects comes

5000 ly

This composite image of Centaurus A shows the elliptical galaxy in optical wavelengths and the double lobe structure of the gas jet in radio, vividly illustrating their close interrelationship.

from a very small region, perhaps only a light year in size, buried in the core of the quasar or active galaxy.

The nearest galaxy with high levels of activity, NGC 5128, lies about 13 million light years from the Earth. This elliptic galaxy has a classic set of double radio lobes much like Cygnus A. The outer lobes are much farther apart in NGC 5128 than in Cygnus A, with each one lying about one million light years from the center. The radio emission from these huge lobes constitutes one of brightest radio sources in the sky: Centaurus A. A much smaller set of inner radio lobes lies about 20,000 light years from the center of the galaxy. By focusing on the smaller lobes, scientists have been able to investigate the phenomenon of double-lobed jets on a smaller physical scale than in Cygnus A.

Jack Burns of New Mexico State University and David Clarke and Michael Norman of the National Center for Supercomputing Applications were able to construct a high-resolution radio image of Centaurus A, using computers at the Very Large Array and the Cray X-MP at NCSA to combine radio observations taken at multiple wavelengths by the Very Large Array.

Burns superimposed that radio emission map over a digitized color optical image of the galaxy to illustrate how closely the jet and its lobes are tied to the elliptical galaxy. The knots of emission along the jet are probably caused by internal shocks as described in Chapter 3. Their visibility in the x-ray region indicates their high energy. Unfortunately, dust lanes running across the galaxy's center obscure the optical structure of the core region where the jet arises.

We can probe the inner region of an active galaxy more closely by looking at the giant elliptical galaxy M87, at the center of the nearby Virgo Cluster of galaxies, 50 million light years from Earth. From optical photographs, it has been known since 1918 that a jet emerges from the inner core of M87. Using the Hubble Space Telescope, Tod Lauer and Roger Lynds of the National Optical Astronomy Observatories and Sandra Faber of the University of California at Santa Cruz and their colleagues were able to create a high-resolution, computer-processed optical image of the very center of this galaxy. The abrupt transverse flat structure in the center of the right half of the image is very similar to a Mach disk like those discussed in Chapter 3.

500 ly

This computer-enhanced Hubble Space Telescope image shows the inner six thousand light-years of the M87 core at optical wavelengths. The image reveals both the likely existence of a central black hole and its probable role in creating the outflowing jet. The rapid increase in brightness toward the core signals the presence of a central cusp of stars surrounding the billion-solar-mass black hole.

The origin of the jet is clearly the very bright central spike. Only an object of about a billion solar masses—a supermassive black hole—could produce a light spike this sharp. The Space Telescope has seen similar light spikes at the center of a number of other spiral and elliptical galaxies; thus, a black hole may commonly dwell in galactic centers. Even our own spiral galaxy shows evidence for a black hole at its center. Because our galactic center is 2000 times closer to Earth than the center of M87, we can resolve its core in much greater detail.

Using NCSA's Cray X-MP supercomputer, University of Illinois astronomers Neil Killeen and Kwok-Yung Lo followed a procedure similar to that described for Burns and Clarke to

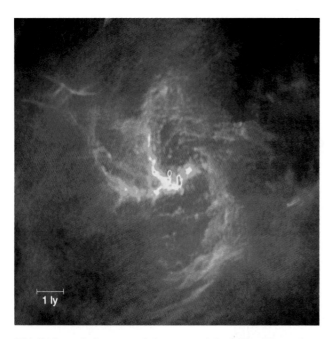

This high-resolution map of the center of the Milky Way galaxy clearly shows a pinwheel of large-scale magnetic field and ionized gas. The intensity of the 6-centimeter-wavelength radio emission is color coded from blue to red in the order of the rainbow.

obtain a very high resolution radio map of our galaxy's center. In the inner ten light years of the central region, there is a pinwheel-shaped complex of ionized gas, spiraling inward toward the center. We can obtain the velocities of clouds and stars in this region from their Doppler shifts. These velocities are consistent with the velocities we would expect for orbital motion around a black hole of a few million solar masses.

At the center of all this activity is a point radio source called Sagittarius A*. Lo and Don Backer of the University of California at Berkeley and their colleagues have shown that this radio source must be smaller in diameter than the orbit of Jupiter around the Sun—an incredibly small size for such a powerful radio source. Furthermore, the radio spectrum of this object resembles that of much stronger radio sources in the cores of other active galaxies, including quasars. This object, unique in the galaxy, may well be a magnetized disk of gas, termed an accretion disk, spiraling around a central black hole.

The observational evidence seems to be overwhelming that supermassive black holes are common at the centers of galaxies and quasars. How did they come to be there? During the formation of galaxies, a large amount of gas probably accumulated at the core. Much of this gas fragmented into clouds, and these clouds then collapsed to form stars. The density at the center of the star cluster became so great that its core in turn started to collapse. Stuart Shapiro and Saul Teukolsky of Cornell University have used thousands of hours on the IBM 3090 at the Cornell National Supercomputer Facility to study in detail the evolution of such star clusters. They do indeed find that under the correct conditions the overly dense cores of star clusters develop such a strong gravitational field that they form a supermassive black hole.

Taken together, these observations and simulations strongly suggest that in active galactic nuclei and quasars the mysterious central engine is a supermassive black hole accreting gaseous matter. The gas falling toward a black

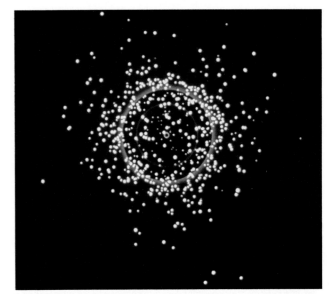

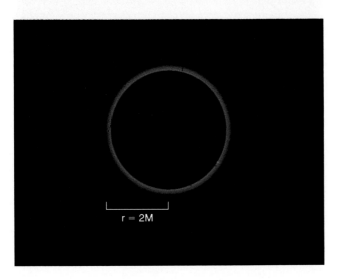

Three stages in the collapse of an unstable star cluster. Initially the stars are far enough apart that light can escape. At a certain moment in the collapse a black hole forms (blue line), trapping light within. Quickly the stars fall into the hole, leaving only the empty space surrounding the black hole. The radius of the hole is two times its mass (2M) in units in which the Newtonian constant G and the speed of light c are set equal to one.

r = 2M

hole should possess some angular momentum, perhaps from the rotation of the surrounding galaxy. The combined effects of gravity and angular momentum cause the gas to settle into a disk or torus centered on the hole.

Supercomputer simulations, one of which we studied in Chapter 2, of accretion onto a black hole were pioneered by James Wilson at Lawrence Livermore National Laboratory, John Hawley at the University of Virginia, and Larry Smarr at the University of Illinois. Although these simulations neglected the effects of magnetic fields, radiation pressure, and other factors, they already show how gas flow onto a black hole naturally organizes itself into an accretion torus and two outflowing jets.

One would think that a central black hole would soon "mop up" all the available gas in its

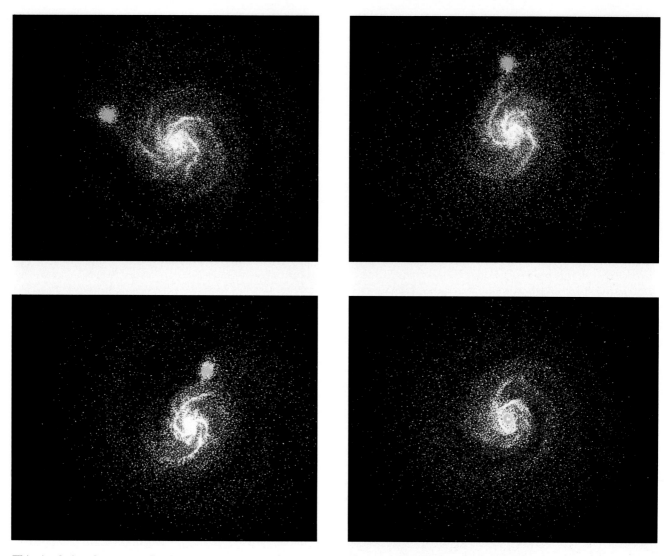

This simulation shows a small galaxy (stars in orange) being devoured by a larger, disk-shaped galaxy. The pictures display progress at 400-million-year intervals.

vicinity and then become dormant. Indeed, most galaxies seem to pass through long periods of low activity, as our own galaxy seems to be in now. What, then, can "rejuvenate" a black hole and lead to a new epoch of activity? To find an answer, we must once again move upward in scale, from individual galaxies to clusters of galaxies. If a galaxy is in the center of a rich galac-

tic cluster, as M87 is in the center of the Virgo cluster, it may capture and devour additional galaxies and grow to enormous dimensions. This phenomenon is called galactic cannibalism. Cannibalism differs from the mergers that we studied in Chapter 3 in that the dining galaxy is larger than its dinner, whereas merging galaxies are about the same size.

Lars Hernquist at the University of California at Santa Cruz has used a Cray Y-MP at the Pittsburgh Supercomputing Center to simulate a large, disk-shaped galaxy devouring a small satellite galaxy. By mass, 90 percent of the large galaxy consists of stars (in blue in the image on this page) and 10 percent of gas (in white). A halo of dark matter surrounds the large galaxy; its mass surpasses that of the disk by about three times. The satellite galaxy, which has a mass a tenth that of the large galaxy, contains only stars (in orange). Initially, the satellite is in circular orbit about its larger companion, then as the collision proceeds it spirals in toward the core of the large galaxy. The effects of gravity cause spiral arms to appear in the large galaxy during the small galaxy's two-billion-year journey to the core. Although much material has been stripped from the satellite, most of its stars plunge into the nucleus of the large galaxy.

As the satellite galaxy sweeps through the inner regions of the disk galaxy during the final stages of this simulation, its passage causes a significant amount of gas to become concentrated along one of the spiral arms. This compressed interstellar gas will eventually cool and clump to form many protostars. The simulation therefore demonstrates that galactic collisions can stimulate prolific star formation, and thus they may account for "starburst galaxies" that blaze with the light of numerous newborn stars. A cannibalistic collision of this type could also "feed" fresh gas to a black hole at the core of the larger galaxy. The large amount of newly infalling matter would release enormous energy, thereby explaining the vastly greater power output of some galaxies and quasars.

The Large-Scale Structure of the Universe

From planets to solar systems, from stars to galaxies, from galaxies to clusters of galaxies, the universe builds to larger scales in a hierarchical fashion. At what level do we reach the highest level of structure in the universe? Perhaps in this decade, the advent of teraflop supercomputers and computer-automated telescopes will bring the answer within our grasp.

Most of the gravitational mass of the universe is not luminous. This dark matter has roughly ten times the mass that exists in the stars that light the cosmos. It can aggregate in clumps as small as galaxies, but it also seems to form structures on scales even larger than clusters of galaxies. The luminous gaseous matter presumably adds little to the large-scale gravitational field of the universe. But because the luminous matter clusters in the powerful gravitational field of the dark matter, it traces out the underlying concentrations of dark matter that creates the universe's large-scale gravitational field.

If we only knew the location in three-dimensional space of every visible galaxy, we would be able to map the gravitational structure of the nearby universe. Since the 1920s, it has been known that galaxies are receding as the universe expands, and that the distance of a galaxy is directly proportional to its velocity of recession. Because of the Doppler effect, the recessional velocity is proportional to the red shift of the spectral lines emitted by the stars in a galaxy. Therefore, measuring a galaxy's redshift gives its distance. Scientists are now building a database of galaxy positions in the sky and their redshift-determined distances from which to construct a three-dimensional map of gravitational structure.

Although errors from many sources can enter the map of gravitational structure, one could correct for most of them if one had measured a large enough ensemble of galaxies. For more than four decades astronomers have been painstakingly measuring a larger and larger sample of galaxy redshifts. By 1990, the roughly 8000 redshifts known in 1980 had grown to over 30,000.

One of the most ambitious and systematic efforts to map galaxies has been proceeding for the past ten years at the Harvard University-Smithsonian Center for Astrophysics under the leadership of Margaret Geller and John Huchra. Geller and her colleagues have measured the individual redshifts of thousands of galaxies that lie within long, thin strips on the sky. When they extend a strip outward to put the galaxies at their appropriate distance from Earth, they obtain thin, wedge-shaped slices, stretching out to nearly a billion light years from Earth.

Their results reveal that galaxies tend to cluster in thin shells surrounding roughly spherical voids of enormous diameter—as great as several hundred million light years. When Huchra and Geller's team combined several of their wedges, they discovered an enormous thin sheet of galaxies, which they call the Great Wall. From these results, they and other researchers infer that galaxies surround spherical voids in the same way that soap film is concentrated on the surface of bubbles.

With the advent of automated telescopes able to measure many redshifts simultaneously,

the number of galaxies in the database should grow to hundreds of thousands by late in the decade. Some experiments are already underway to develop more sophisticated interactive visualization techniques to help scientists analyze these large data sets. For instance, Geller and her colleague Emilio Falco, working with graphics specialists Michael McNeill, Mark Bajuk, and Mike Krogh, were recently able to explore their sample with NCSA's virtual reality viewing devices. These devices allowed the scientists to view their galaxy maps as though they were moving through the maps at will, taking any path, at any angle or range they chose.

As the structure of the nearby universe is coming into focus observationally, efforts have intensified to understand how this gravitational structure came to be. This large-scale structure may reasonably result from primordial conditions that date back to the earliest moments of the universe. We can see into the past by observing objects so distant that their light is just reaching us now after traveling millions or even

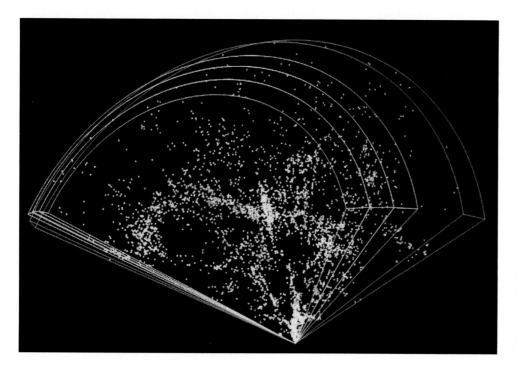

This map shows the distribution of 4000 galaxies in four slices. The "Great Wall" crosses the survey nearly parallel to the outer boundary. Earth is at the lower tip of all the wedges.

Margaret Geller uses the Fakespace BOOM virtual reality gear to view one slice of the data. In the background, the image she is inspecting has been projected as blue and red images for stereo viewing by a group.

billions of light years. We recognize the distance traveled and the corresponding amount of time elapsed by the value of the redshift. The redshifts of nearby galaxies are shifted only a few percent from their "at rest" wavelengths. Even the most distant quasars have a redshift of 5, presumably marking the time in the past when galaxies were beginning to form.

Beyond that point there is nothing observable until a red shift of 1000. At this moment, the universe, which today is about 15 billion years old, was only about one million years old. Light coming from that distant time forms the ubiquitous microwave background radiation, the relic of the ancient time when photons and matter were in close contact. Any inhomogeneities in the distribution of matter at that early time would show up as distortions in the background radiation. The recent observations of the Cosmic Background Explorer satellite have shown that the microwave background is smooth to a few parts in 100,000.

The near-isotropy of the microwave background tells us that the primordial fireball was a remarkably homogeneous mixture of particles and radiation. But galaxies could not have formed unless there were some slight deviations from perfect uniformity. These deviations probably began as small, long-wavelength perturbations in the density of matter that eventually grew by the action of gravity to become the enormous, sudsy structures that we see in the universe today.

Coincidentally, supercomputers have recently developed memories large enough for these machines to be used to model structure at about this same scale. As discussed in Chapter 3, simulations of the evolution of dark matter using particle-mesh techniques did find filaments and voids forming. However, the early pioneering work of Centrella and Melott was barely able to resolve the dark matter distribution, because of the small memory of the Cray X-MP. More recent work by Edmund Bertschinger and James Gelb of MIT used the much larger memory and the vector-parallel architecture of the IBM 3090 at the Cornell National Supercomputer Facility to run a grid six times finer in each spatial dimension.

Their simulations are able to follow more than one million particles representing the dark matter. A galaxy the size of our Milky Way

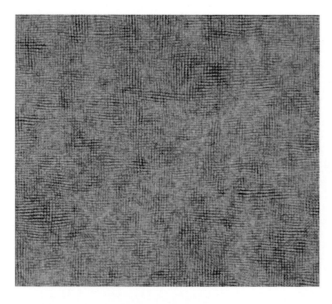

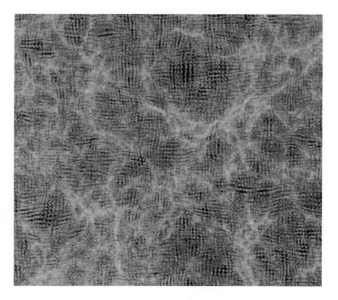

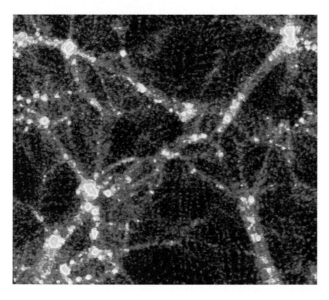

Slices through a large volume of space show the formation of large-scale structure in the distribution of dark matter in the universe. When the universe was much smaller, only small deviations existed (top left). Just before galaxy formation, significant clumping can be seen (top right). At today's epoch, the dark matter is clustered in filaments and shells surrounding more empty voids (lower left).

would be represented by about 1000 particles, small compared to the 10 billion stars in the galaxy, but enough to give unprecedented resolution. This code can even begin to resolve instances of galactic cannabalism like those studied in fine detail for individual galaxies by Hernquist. High resolution is important, because it allows us to follow how the hills and valleys in the gravitational field are formed by the increasingly inhomogeneous distribution of the dark matter.

The last stage in simulating the evolution of large-scale structure is to combine gas dynamics codes, to follow the luminous matter, with particle codes, to follow the dark matter. Unlike the collisionless dark matter, the gaseous

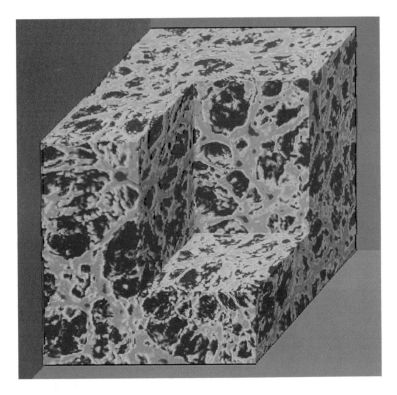

A volume visualization from a supercomputer simulation of the density of the luminous matter coupled to the dark matter in the universe. One can see the higher density knots of matter (red) forming at the intersection of the filaments (yellow). The lower density voids are in blue.

matter can form shocks that will then alter the flow of gas. The shock-heated matter can radiate its energy away, perhaps forming some of the observed diffuse background of x-rays. Such hybrid codes are being developed by Jeremiah Ostriker and Renyue Cen of Princeton University in collaboration with Michael Norman of the National Center for Supercomputing Applications. Using the large-memory Convex supercomputer at NCSA, they have already made simulations that follow the evolution of both luminous and dark matter. They find a similar structure of filaments and voids in both types of matter.

However, their code is still not adequate to test against the details of galaxy distribution. The critical physics is occurring on spatial scales ranging over many orders of magnitude. The phenomena on these scales must each be resolved and coupled to the others by exploiting new techniques in adaptive mesh refinement. New algorithms must be developed to make

both particle and gas dynamics codes run efficiently on massively parallel supercomputers. No one scientist can have deep enough knowledge in each of the areas of science and computing techniques required to attack such an ambitious problem. Therefore, teams of experts are beginning to form, as called for in the High Performance Computing and Communications Program described in Chapter 1, to mount a multiyear assault. These teams require experts in gas dynamics, N-body codes, and the astrophysical processes, as well as in the new areas of algorithm design required to run multiscale problems on massively parallel machines.

The efforts of such teams will bring forth, over the next five years, nationally accessible computer programs for simulating the formation of the universe. The work of theorists, observers, experimentalists, and computational scientists will merge as all seek to understand more deeply the structure of the universe.

Further Readings

For the reader who wishes to explore further examples of the use of supercomputers in specific research projects, there are several magazines that often cover supercomputing applications: *Computers in Physics, Cray Channels, Nature, Science, Science News, Scientific American,* and *Supercomputing Review,* now *Supernet.* In addition, most supercomputer centers and their funding agencies have annual reports on science and engineering applications. The NSF centers also publish newsletters featuring recent applications carried out at their facilities:

Access. National Center for Supercomputing Applications, University of Illinois, Urbana, Ill., 61801

Gather/Scatter. San Diego Supercomputer Center, P.O. Box 85608, San Diego, Calif. 92186-9784

Forefronts. Cornell Theory Center, Cornell University, Ithaca, N.Y. 14853-3801

PSC NEWS. Pittsburgh Supercomputing Center, Carnegie-Mellon University, Pittsburgh, Pa. 15213

In the list of suggested readings that follows, books are listed according to the order of topics within the chapter. Many of the books suggested for Chapters 4 through 8 do not discuss supercomputing specifically, but rather give general background in the subject areas.

Chapter 1

S. Karin, and N. P. Smith. *The Supercomputer Era.* Orlando, Fla.: Harcourt Brace Jovanovich, 1987.

C. Lazou. *Supercomputers and Their Use,* rev. ed. Oxford: Clarendon Press, 1988.

R. M. Friedhoff and W. Benzon. *Visualization: The Second Computer Revolution.* New York: Freeman, 1991.

R. B. Wilhelmson. *High-Speed Computing: Scientific Applications and Algorithm Design.* Urbana: University of Illinois Press, 1988.

E. J. Pitcher, ed. *Science and Engineering on Supercomputers.* Proceedings of the Fifth International Conference. Southampton: Computational Mechanics Publications; Berlin: Springer-Verlag, 1990.

E. J. Pitcher, ed. *Science and Engineering on Supercomputers.* Proceedings of the Fourth International Conference. Minneapolis: Cray Research, 1989.

E. J. Pitcher, ed. *Science and Engineering on Supercomputers.* Proceedings of the Third International Conference. Minneapolis: Cray Research, 1988.

H. D. Simon, ed. *Scientific Applications of the Connection Machine.* Singapore: World Scientific, 1989.

J. Mesirov, ed. *Very Large Scale Computation in the 21st Century.* Philadelphia: Society for Industrial and Applied Mathematics, 1991.

Edwin Kessler, ed. *Thunderstorm Morphology and Dynamics.* 2d ed. Norman, Okla.: University of Oklahoma Press, 1986.

H. Rheingold. *Virtual Reality.* New York: Simon & Schuster, 1991.

B. Kahin. *Building Information Infrastructure.* New York: McGraw-Hill, 1992.

J. R. Kirkland and J. H. Poore, eds. *Supercomputers: A Key to U.S. Scientific, Technological, and Industrial Preeminence.* New York: Praeger Publishers, 1987.

National Research Council. *The National Challenge in Computer Science and Technology.* Washington, D.C.: National Academy Press, 1988.

National Research Council. *Computing Our Future.* Washington, D.C.: National Academy Press, 1992.

Federal Coordinating Council for Science, Engineering, and Technology. Office of Science and Technology Policy. *Grand Challenges 1993: High Performance Computing and Communications.* Washington, D.C.

Chapter 2

S. Augarten. *Bit by Bit: An Illustrated History of Computers.* New York: Ticknor & Fields, 1984.

Time-Life Books, eds. *Understanding Computers: Speed and Power.* Alexandria, Va.: Time-Life Books, 1990.

Time-Life Books, eds. *Understanding Computers: Illustrated Chronology.* Alexandria, Va.: Time-Life Books, 1990.

T. R. Reid. *The Chip.* New York: Simon & Schuster, 1984.

H. Rheingold. *Tools for Thought.* New York: Prentice-Hall, 1986.

R. M. Hord. *The ILLIAC IV: The First Supercomputer.* Rockville, Md.: Computer Science Press, 1982.

J. L. Hennessy and David A. Patterson. *Computer Architecture: A Quantitative Approach.* San Mateo, Calif.: Morgan Kaufmann, 1990.

R. A. Jenkins. *Supercomputers of Today and Tomorrow: The Parallel Processing Revolution.* Blue Ridge Summit, Pa.: Tab Books, 1986.

K. M. Chandy and S. Taylor. *An Introduction to Parallel Programming.* Boston: Jones and Bartlett, 1992.

National Research Council. *Supercomputers: Directions in Technology and Applications.* Washington, D.C.: National Academy Press, 1989.

W. D. Hillis. *The Connection Machine.* Cambridge, Mass.: MIT Press, 1986.

Chapter 3

R. L. Bowers and J. R. Wilson. *Numerical Modeling in Applied Physics and Astrophysics.* Boston: Jones and Barlett, 1991.

D. S. Burnett. *Finite Element Analysis.* Boston: Addison-Wesley, 1988.

H. Kardestuncer and D. H. Norrie. *Finite Element Handbook.* New York: McGraw-Hill, 1987.

R. W. Hockney and J. W. Eastwood. *Computer Simulation Using Particles.* New York: McGraw-Hill, 1981.

W. H. Press, B. P. Flannery, S. A. Teukolsky, and W. T. Vetterling. *Numerical Recipes: The Art of Scientific Computing.* Cambridge: Cambridge University Press, 1989.

W. J. Minkowycz, E. M. Sparrow, G. E. Schneider, and R. H. Pletcher. *Handbook of Numerical Heat Transfer.* New York: John Wiley, 1988.

M. P. Allen and D. J. Tildesley. *Computer Simulation of Liquids.* Oxford: Oxford University Press, 1989.

Chapter 4

National Research Council. *Physics Through the 1990s: Elementary-Particle Physics.* Washington, D.C.: National Academy Press, 1986.

National Research Council. *Physics Through the 1990s: Nuclear Physics.* Washington, D.C.: National Academy Press, 1986.

National Research Council. *Physics Through the 1990s: Condensed-Matter Physics.* Washington, D.C.: National Academy Press, 1986.

National Research Council. *Materials Science and Engineering for the 1990s.* Washington, D.C.: National Academy Press, 1989.

National Research Council. *The Impact of Supercomputing Capabilities on U.S. Materials Science and Technology.* Washington, D.C.: National Academy Press, 1988.

Journal of Computational Physics. Academic Press.

K. E. Drexler and C. Peterson with G. Pergamit. *Unbounding the Future: The Nanotechnology Revolution.* New York: William Morrow, 1991.

K. Hess, J. P. Leburton, and U. Ravaioli. *Computational Electronics: Semiconductor Transport and Device Simulation.* Boston: Kluwer Academic Publishers, 1991.

National Research Council. *Opportunities in Chemistry.* Washington, D.C.: National Academy Press, 1985.

P. W. Atkins. *Physical Chemistry.* 4th ed. New York: W. H. Freeman, 1990.

A. Szabo. *Modern Quantum Chemistry: Introduction to Advanced Electronic Structure Theory.* New York: McGraw-Hill, 1989.

T. Clark. *A Handbook of Computational Chemistry: A Practical Guide to Chemical Structure and Energy Calculations.* New York: John Wiley, 1985.

Chapter 5

The Molecules of Life: Readings from Scientific American. New York: W. H. Freeman, 1985.

J. Darnell, H. Lodish, and D. Baltimore. *Molecular Cell Biology.* 2d ed. New York: Scientific American Books, 1990.

C. Branden and J. Tooze. *Introduction to Protein Structure.* New York: Garland Publishing, 1991.

J. A. McCammon and S. C. Harvey. *Dynamics of Proteins and Nucleic Acids.* Cambridge: Cambridge University Press, 1987.

S. Grossberg, ed. *Neural Networks and Natural Intelligence.* Cambridge, Mass.: MIT Press, 1988.

Chapter 6

O. C. Zeinkiewicz and R. L. Taylor. *The Finite Element Handbook.* 4th ed. New York: McGraw-Hill, 1989.

A. K. Noor and J. T. Oden. *State-of-the-art Surveys on Computational Mechanics.* New York: American Society of Mechanical Engineers, 1989.

D. A. Anderson, J. C. Tannehill, and R. H. Pletcher. *Computational Fluid Mechanics and Heat Transfer.* Bristol, Pa.: Hemisphere Publishing, 1984.

Chapter 7

F. Press and R. Siever. *Earth.* 4th ed. New York: W. H. Freeman, 1986.

S. M. Stanley. *Earth and Life Through Time.* 2d ed. New York: W. H. Freeman, 1988.

W. Washington and C. L. Parkinson. *An Introduction to Three-Dimensional Climate Modeling.* Mill Valley, Calif.: University Science Books, 1986.

S. H. Schneider. *Global Warming: Are We Entering the Greenhouse Century?* San Francisco: Sierra Club Books, 1989.

R. Costanza, ed. *Ecological Economics: The Science and Management of Sustainability.* New York: Columbia University Press, 1991.

National Research Council. *Toward an Understanding of Global Change.* Washington, D.C.: National Academy Press, 1988.

National Research Council. *Global Change and Our Common Future.* Washington, D.C.: National Academy Press, 1989.

Chapter 8

National Research Council. *The Decade of Discovery in Astronomy and Astrophysics.* Washington, D.C.: National Academy Press, 1991.

W. J. Kaufmann III. *Universe.* 3d ed. New York: W. H. Freeman, 1991.

J. Centrella, J. LeBlanc, and R. Bowers. *Numerical Astrophysics.* Boston: Jones and Bartlett, 1985.

Journal of Computational Astrophysics. San Diego: Academic Press.

Sources of Illustrations

Drawings by Ian Worpole and Vantage Art

Frontispiece
Grafik Communications

Facing page 1
Matthew Arrott, Colleen Bushell, Mark Bajuk, Jeffrey Thingvold, Jeffrey Yost, Robert Wilhelmson, Brian Jewett, Louis Wicker, NCSA

Page 3
Cray Research, Inc.

Page 9
Paul Woodward, Minnesota Supercomputing Institute

Page 10
Adapted from a drawing by H. B. Bluestein and C. R. Parks. *Monthly Weather Review*, Vol. 111, 1983, p. 2034.

Page 11
Left, Peter Ray, Florida State University; Robert Wilhelmson, University of Illinois, Urbana-Champaign; and Ken Johnson, Florida State University. *Bulletin for the American Meteorological Society*, Vol. 3, No. 3, March, 1982, cover
Right, Joseph Klemp, National Center for Atmospheric Research

Page 12
Robert Wilhelmson, University of Illinois, Urbana-Champaign. *Journal of Atmospheric Science*, Vol. 38, 1981, p. 1581.

Page 13
Stefen Fangmeier, Robert Wilhelmson, Harold Brooks, Louis Wicker, Crystal Shaw, NCSA

Page 14
Colleen Bushell, Matthew Arrott, Polly Baker, Michael McNeill, NCSA; Edward Tufte

Page 15
Matthew Arrott, Colleen

Bushell, Mark Bajuk, Jeffrey Thingvold, Jeffrey Yost, Robert Wilhelmson, Louis Wicker, NCSA

Page 16
Matthew Arrott, Colleen Bushell, Mark Bajuk, Jeffrey Thingvold, Jeffrey Yost, Robert Wilhelmson, Louis Wicker, NCSA

Page 21
Robert Patterson and Donna Cox, NCSA

Page 24
Thinking Machines Corporation, Inc.

Page 26
David Exton, Science Museum, London

Page 31
University of Pennsylvania

Page 40
Lawrence Livermore National Laboratory

Page 42
Left, NCSA
Right, Cray Research, Inc.

Page 43
Cray Research, Inc.

Page 44
John Hawley, University of Virginia, and Larry Smarr, NCSA

Page 45
D. H. Thompson, Aeronautical Research Laboratories, Defense Science and Technical Organization

Page 46
Kozo Fujii and Lewis Schiff, NASA Ames

Page 47
Adapted from a graph created by Steve Wallach, Convex Computer Corporation.

Page 48
nCUBE, Inc.

Page 51
National Science Foundation Supercomputer Metacenter

Page 53
Adapted from Elizabeth Corcoran, "Calculating Reality," *Scientific American*, Vol. 264., No. 1, January 1991, p. 109

Page 54
Erich Wimmer, Cray Research, and Arthur Freeman, Northwestern University

Page 58
Adapted from a graph in R. D. Richtmyer and K. W. Morton, *Difference Methods for Initial-Value Problems*, Interscience: 1967

Page 59
Adapted from a graph in R. D. Richtmyer and K. W. Morton, *Difference Methods for Initial-Value Problems*, Interscience: 1967

Page 60
Larry Morris, Washington Post

Page 61
Top, NASA
Bottom, National Radio Astronomy Observatory; Observers R. A. Perley and J. W. Dreher

Page 62
Michael Norman, NCSA

Page 63
Michael Norman, NCSA

Page 64
Michael Norman, NCSA

Page 65
Michael Norman, NCSA

Page 66
Karl-Heinz Winkler, Los Alamos National Laboratory

Page 67
Top, Michael Norman, NCSA; Phil Hardee, University of Alabama; Donna Cox, NCSA
Bottom, Michael Norman, NCSA

Page 69
Top, adapted from a figure in J. N. Reddy, *An Introduction to the Finite Element Method*, McGraw-Hill: 1984
Bottom, adapted from a graph in J. N. Reddy, *An Introduction to the Finite Element Method*, McGraw-Hill: 1984

Page 70
Simulation of the flow past a Dassault Aviation Falcon computed with the code FLITE. Data courtesy of J. Peiro, J. Peraire, and K. Morgan, Imperial College, London, UK

Page 71
Robert Haber, University of Illinois at Urbana-Champaign; adapted from Eric Pitcher, ed., *Science and Engineering on Cray Supercomputers*, Minneapolis: Cray Research, pp. 141–142

Page 72
Robert Haber, NCSA

Page 73
Robert Haber, University of Illinois at Urbana-Champaign; adapted from Eric Pitcher, ed., *Science and Engineering on Cray Supercomputers*, Minneapolis: Cray Research, pp. 141–142

Page 74
Palomar Observatory

Page 75
Joshua Barnes, Institute for Astronomy, Hawaii

Page 76
Joan Centrella, Drexel University

Page 79
David Ceperley. *access*, November-December, 1988, p. 5

Page 82
Jeffrey Thingvold, NCSA; Aileen Alvardo-Swaisgood, Amoco Oil Company

Page 84
William Sherman, Jeffrey Yost, Jeffrey Thingvold, NCSA; David Herron, Eli Lilly and Company

Page 86
Jerzy Bernholc, North Carolina State University. *Science*, Vol. 254, 20 December, 1991, cover

Page 89
Matthew Arrot, NCSA; NHK.

Page 91
Thomas DeGrand, University of Colorado

Page 92
Adapted from a graph by Gordon Baym. *Physics Today*, March 1985, p. 3

Page 93
Images produced at the Pittsburgh Supercomputing Center by P. K. Hsiung, Robert Dunn, and Nathan Loofbourrow of Carnegie Mellon University in collaboration with Joel Welling of the Pittsburgh Supercomputing Center computer graphics staff.

Page 94
James Wilson and Thomas McAbee, Lawrence Livermore National Laboratory

Page 97
Adapted from a graph created by William Magro and David Ceperley, University of Illinois at Urbana-Champaign.

Page 98
Troy Barbee III, Marvin L. Cohen, University of California, Berkeley. *Physical Review*

B, Vol. 44, No. 21, 1991, p. 565

Page 99
IBM

Page 100
Erich Wimmer, Cray Research, and Arthur Freeman, Northwestern University

Page 101
Ronald Cohen, Carnegie Institution of Washington. *Science*, Vol. 255, 3 January 1992, cover.

Page 103
Adapted from an image by Karl Hess. *Physics Today*, February 1990, p. 41

Page 105
Animations were created from supercomputer simulations by Umberto Ravaioli and L. Frank Register of the Beckman Institute at the University of Illinois at Urbana-Champaign

Page 106
Adapted from Konstantin K. Likharev and Tord Claeson, "Single Electrons," *Scientific American*, Vol. 266, No. 6, June 1992, p. 85

Page 107
Joseph W. Lyding, Beckman Institute, University of Illinois at Urbana-Champaign

Page 108
Top, IBM
Bottom, Joseph W. Lyding, Beckman Institute, University of Illinois at Urbana-Champaign

Page 109
Uzi Landman, W. D. Leudtke, N. A. Burnham, and R. J. Colton, Georgia Institute of Technology.

Page 110
Harrell Sellers, NCSA

Page 112
David Dixon, E. I. Du Pont de Nemours & Company

Page 113
David Dixon, E. I. Du Pont de Nemours & Company

Page 114
Top, Paul Bash, Martin Field, and Martin Karplus, Harvard University; Matthew Arrott, Michelle Mercer, Jeffrey Yost, NCSA
Bottom, Paul Bash and Martin Karplus, Harvard University, and Robert Davenport of MIT; Matthew Arrott, Michelle Mercer, Jeffrey Yost, NCSA

Page 116
Ray-traced image by Andreas Windemuth from a simulation by Helmut Keller, Michael Schaefer, and Klaus Schulten, Beckman Institute, University of Illinois at Urbana-Champaign

Page 119
Bob Hermann, Eli Lilly and Company

Page 121
Carl Woese and Gary Olson, University of Illinois at Urbana-Champaign

Page 123
Susan Taylor, University of California at San Diego; a slightly different version appeared on the cover of *Science*, Vol. 23, July 26, 1991.

Page 124
Bob Hermann, Eli Lilly and Company

Page 125
Richard Goldstein, Peter Wolynes, and Zan Schulten, University of Illinois at Urbana-Champaign

Page 126
Ron Elber, University of Illinois at Chicago

Page 127
John Rosenberg, University of Pittsburgh

Page 128
U. Chandra Singh and Frederick H. Hausheer, BioNumerik Pharmaceuticals

Page 129
Preparation of capsids and antibodies: W. W. Newcomb and J. C. Brown, University of Virginia; structural analysis and

computer graphics: F. P. Booy, J. F. Conway, A. C. Steven, and B. L. Trus, NIH; computer 3-D software: T. Baker, Purdue University, and C. Johnson, NIH; computer systems for 3-D reconstruction hardware: DEC 4000-300 and Intel iPSC/860

Page 130
Michael Rossmann, Purdue University; *Science*, Vol. 233, No. 4770, September 9, 1986, p. 1286.

Page 131
X-ray structure solved by members of the Crystallography Laboratory, National Cancer Institute using synthetic protein supplied by Caltech; computer graphics image produced by the Computer Graphics Laboratory, University of California, San Francisco, directed by Robert Langridge

Page 132
Shankar Subramaniam. *Journal of Immunology*, Vol. 146, 1991, p. 4255

Page 133
Art Olson, Scripps Institute

Page 135
Shankar Subramaniam, Eric Jakobsson, and See-Wing Chiu, NCSA and the University of Illinois at Champaign

Page 136
M. E. Clark, University of Illinois at Urbana-Champaign; NCSA

Page 137
M. E. Clark, University of Illinois at Urbana-Champaign; NCSA

Page 138
Data supplied by Eric Hoffman, University of Pennsylvania; image by Pat Moran, Clint Potter, NCSA

Page 139
Charles S. Peskin, David M. McQueen, Courant Institute of Mathematical Sciences, New York University

Page 140
Paul Lauterbur, Beckman Institute, University of Illinois at Urbana-Champaign; Clint Potter, NCSA

Page 141
Mark Ellisman, David Hessler, and Stephen Young of the San Diego Microscopy and Imaging Resource supported by the National Center for Research Resources of the NIH; Steve Lamont, San Diego Supercomputer Center

Page 142
Klaus Obermayer, Helge Ritter, and Klaus Schulten, Beckman Institute, University of Illinois at Urbana-Champaign; Gary Blasdel, Harvard Medical School

Page 144
Mark Bajuk, NCSA; Rich Ingram, Caterpillar, Inc.

Page 146
Glen Villa, MacNeal Schwendler Corporation for MacGregor Golf Company. *Supercomputing Review*, August 1991, p. 32

Page 147
Michael McNeill, NCSA; Kishore Kar, Dow Chemical Company

Page 149
Jonathan Dantzig, University of Illinois at Urbana-Champaign; Cray Research, Inc.

Page 150
Pittsburgh Supercomputing Center

Page 151
Ashwini Kumar, Southeast Dairy Foods Research Center

Page 152
FMC Corporation

Page 153
FMC Corporation

Page 154
T. J. Hanratty, and S. L. Lyons, University of Illinois at Urbana-Champaign

Page 155
S. K. Robinson, NASA Langley

Research Center; S. J. Kline, Stanford University; P. R. Spalart, Boeing Commercial Aircraft Company

Page 156
David Gosman, Computational Dynamics, Ltd., for creation of mesh and performance of the analysis; STAR-CO is a product of Computational Dynamics, Ltd., London, and was used for calculations

Page 157
Top, Advanced Industrial Concepts Division, DOE, and the Fluid Dynamics Group, T-3, Los Alamos National Laboratories
Bottom, Susumu Shirayama, Softek Systems Inc., Japan

Page 158
A. Guillet, Centre d'Etudes et de Recherche de Ladoux, Michelin Corporation. *Cray Channels*, Winter 1989, p. 13

Page 159
K. Gruber and T. Frank, Mercedez-Benz AG. *Cray Channels*, Winter 1992, p. 4

Page 160
A. Tilakasiri, Research and Engineering Centre, Ford Motor Company, England, and P. DuBois; in E. J. Pitcher, ed., *Science and Engineering on Supercomputers*, Southampton: Computational Mechanics Publications; Berlin: Springer-Verlag, p. 578.

Page 161
San Diego Supercomputing Center

Page 162
Ralf Gruber, Centre de Recherches en Physique des Plasmas, Ecole Polytechnique Federal de Lausanne, Switzerland; Max Planck Institute fur Plasmaphysik, Garching bei Munchen, Germany

Page 164
Morton Thiokol Inc. Graphics by W. M. Remus. *Cray Channels*, Spring 1988, p. 9.

Page 165
NASA

Page 166
NASA Ames Research Center

Page 168
NOAA/NESDIS/NCDC/SDSD

Page 171
David Bercovici, University of Hawaii. *Science*, Vol. 244, May 26, 1989, p. 953.

Page 172
Craig Upson, NCSA; Craig Bethke, University of Illinois at Urbana-Champaign

Page 173
NCSA

Page 174
Robert Chervin, National Center for Atmospheric Research

Page 175
Robert Chervin, National Center for Atmospheric Research

Page 176
Bill Holland and Frank Bryan, NCAR; image modified by Michael Norman and Larry Smarr, NCSA

Page 177
Illustration by Altf Lannerback, *Dagens Nyheter*, Stockholm

Page 178
David Williamson, University Corporation for Atmospheric Research/National Center for Atmospheric Research/National Science Foundation

Page 179
European Centre for Medium-Range Weather Forecasts

Page 180
European Centre for Medium-Range Weather Forecasts

Page 182
Warren M. Washington, National Center for Atmospheric Research; Jeffrey Yost, NCSA

Page 184
U. Cubasch, Max Planck Institute for Meteorology

Page 187
John Kutzbach, Center for Cli-matic Research, University of Wisconsin–Madison

Page 188
J. Pudykiewicz, Atmospheric Environment Service, Canada

Page 190
A. G. Russell and G. J. McRae, Carnegie Mellon University; Michael McNeill, William Sherman, Mark Bajuk, NCSA

Page 191
Allan Craig, NCSA; David Kovacic, University of Illinois at Urbana-Champaign

Page 192
Fred Sklar, University of South Carolina; Robert Costanza, University of Maryland; Mary White, Louisiana State University

Page 193
Fred Sklar, University of South Carolina; Robert Costanza, University of Maryland; Mary White, Louisiana State University

Page 196
David Arnett, University of Arizona

Page 199
M. E. Kipp, Sandia National Laboratories

Page 200
W. Benz, Harvard-Smithsonian Center for Astrophysics

Page 201
Jet Propulsion Laboratory, CIT, NASA

Page 202
J. A. Fedder and C. M. Mobarry, Naval Research Laboratory; J. G. Lyon, University of Iowa

Page 203
Robert Wolff, Apple Computer; Michael Norman, NCSA

Page 206
D. H. Porter, P. R. Woodward, and Q. Mei, "Simulation of Compressible Convection with the Piecewise-Parabolic Method (PPM)," *Video Journal of Engineering Research*, Vol. 1, No. 1, pp. 1–24

Page 207
Andrea Malagoli and Fausto Cattaneo, University of Chicago

Page 208
Adapted from an illustration by John Maduell, Lawrence Livermore National Laboratory. Stan Woosley and Tom Weaver, "The Great Supernova of 1987," *Scientific American*, Vol. 261, No. 2, August 1989, p. 35

Page 210
National Radio Astronomy Observatory; Observers P. A. Angerhofer, R. Braun, S. F. Gull, R. A. Perley, R. J. Tuffs

Page 211
Mordecai-Mark MacLow, University of Chicago; Richard McCray, University of Colorado; Michael Norman, NCSA

Page 212
Charles Evans, University of North Carolina; Ray Idaszak, North Carolina Supercomputing Center; and NCSA

Page 213
David Hobill, David Bernstein, Larry Smarr, Donna Cox, Ray Idaszak, NCSA

Page 214
Jack O. Burns, New Mexico State University; David Clarke, NCSA

Page 215
Space Telescope Science Institute/AURA/NASA/ESA

Page 216
Kwok-Yung Lo and Neil Killeen, University of Illinois at Urbana-Champaign

Page 217
Stuart L. Shapiro and Saul Teukolsky, Cornell University. *Science*, Vol. 241, 22 July 1988, p. 421

Page 218
Lars Hernquist, University of California, Santa Cruz

Page 220
Margaret Geller, Harvard-Smithsonian Astrophysical Observatory

Page 221
Margaret Geller and Emilio Falco, Harvard-Smithsonian Center for Astrophysics; Michael McNeill, Mark Bajuk, Mike Krogh, NCSA. The software used by the boom was developed by Steve Bryson and Creon Levitt of the NASA Ames Research Center

Page 222
Edmund Bertschinger and James Gelb, MIT

Page 223
Michael Norman, NCSA; Renyue Chen and Jeremiah P. Ostriker, Princeton University

Index

233

Other books in the Scientific American Library Series

POWERS OF TEN
by Philip and Phylis Morrison and the Office of
Charles and Ray Eames

HUMAN DIVERSITY
by Richard Lewontin

THE DISCOVERY OF SUBATOMIC PARTICLES
by Steven Weinberg

FOSSILS AND THE HISTORY OF LIFE
by George Gaylord Simpson

ON SIZE AND LIFE
by Thomas A. McMahon and John Tyler Bonner

THE SECOND LAW
by P. W. Atkins

THE LIVING CELL, VOLUMES I AND II
by Christian de Duve

MATHEMATICS AND OPTIMAL FORM
by Stefan Hildebrandt and Anthony Tromba

FIRE
by John W. Lyons

SUN AND EARTH
by Herbert Friedman

ISLANDS
by H. William Menard

DRUGS AND THE BRAIN
by Solomon H. Snyder

THE TIMING OF BIOLOGICAL CLOCKS
by Arthur T. Winfree

EXTINCTION
by Steven M. Stanley

MOLECULES
by P. W. Atkins

EYE, BRAIN, AND VISION
by David H. Hubel

THE SCIENCE OF STRUCTURES AND
MATERIALS
by J. E. Gordon

THE HONEY BEE
by James L. Gould and Carol Grant Gould

ANIMAL NAVIGATION
by Talbot H. Waterman

SLEEP
by J. Allan Hobson

FROM QUARKS TO THE COSMOS
by Leon M. Lederman and David N. Schramm

SEXUAL SELECTION
by James L. Gould and Carl Grant Gould

THE NEW ARCHAEOLOGY AND THE
ANCIENT MAYA
by Jeremy A. Sabloff

A JOURNEY INTO GRAVITY AND SPACETIME
by John Archibald Wheeler

SIGNALS
by John R. Pierce and A. Michael Noll

BEYOND THE THIRD DIMENSION
by Thomas F. Banchoff

DISCOVERING ENZYMES
by David Dressler and Huntington Potter

THE SCIENCE OF WORDS
by George A. Miller

ATOMS, ELECTRONS, AND CHANGE
by P. W. Atkins

VIRUSES
by Arnold J. Levine

DIVERSITY AND THE TROPICAL
RAIN FOREST
by John Terborgh

STARS
by James B. Kaler

EXPLORING BIOMECHANICS
by R. McNeill Alexander

CHEMICAL COMMUNICATION
by William C. Agosta

GENES AND THE BIOLOGY OF CANCER
by Harold E. Varmus and Robert A. Weinberg